WILLIAM DILLER MATTHEW

Paleontologist

WILLIAM DILLER MATTHEW

Paleontologist

THE SPLENDID DRAMA OBSERVED

Edwin H. Colbert

COLUMBIA UNIVERSITY PRESS   NEW YORK

COLUMBIA UNIVERSITY PRESS

NEW YORK   OXFORD

Copyright © 1992 Columbia University Press
All rights reserved

Library of Congress Cataloging-in-Publication Data

Colbert, Edwin Harris, 1905–
    William Diller Matthew, paleontologist : the splendid drama observed / Edwin H. Colbert.
        p.   cm.
    Includes bibliographical references and index.
    ISBN 0–231–07964–8
    1. Matthew, William Diller, 1871–1930.   2. Paleontologists—Canada—Biography.   3.
Paleontologists—United States—Biography.
I. Title.
QE707.M417C65   1992
560'.9—dc20
[B]   92-14027
                                                                                                                  CIP

Casebound editions of
Columbia University Press books
are Smyth-sewn and printed on permanent
and durable acid-free paper.

Figures 1, 2, 3, 5, 13, 14, 19, 20, 29, 36, 37, 39, 40, 41, 44, 46, 47, 57, 59, 60, and 61 were
provided from family archives by Margaret Matthew Colbert.

Figure 4 was supplied by permission of The Museum of New Brunswick.

Figures 34 and 38 were provided by William Pomeroy Matthew.

Figures 6, 15, and 35 are from photographs taken by Edwin H. Colbert.

Figures 10, 11, 12, 16, 17, 21, 22, 23, 24, 25, 26, 27, 28, 30, 31, 32, 33, 42, 43, 45, 48, 49, 50,
51, 52, 53, 54, 55, 56, and 58 come from the photographic archives of the American Museum of
Natural History, New York City; permission for their use has been granted by the Museum.

Figures 7 and 9 come from the archives of Columbia University; permission for their use has
been granted by the University.

Figures 8 and 18 are from letters in the archives of the New Brunswick Museum; access to
these letters was provided by Dr. Randall F. Miller and Carol Rosevear of the Museum.

Bibliography of William Diller Matthew's work, pp. 201–223, and figure 2 from
"Climate and Evolution" by William Diller Matthew, special publication of the New York
Academy of Science, vol. 1, 1939, reproduced by permission of the Annals of the New York
Academy of Sciences.

Book design: Teresa Bonner
Printed in the United States of America

c 10 9 8 7 6 5 4 3 2 1

The story of life on the earth is a splendid drama.
       —*William Diller Matthew, 1928*

CONTENTS

PREFACE

If the story of William Diller Matthew had been written a half century ago numerous people—members of the family, colleagues, friends—who knew him would have been able to contribute valuable information concerning his life. Today family members and others who were personally acquainted with him are few in number. Therefore the preparation of a biography of this very important scientist has been a matter of consulting archives and personal letters and, especially, of persistent questioning of his daughter (and my wife), Margaret Matthew Colbert.

I married into the Matthew family in 1933, three years after Dr. Matthew's death. At that time and for several decades afterward it was my privilege and pleasure to become acquainted with quite a group of Matthews; besides Margaret there were her sister, Elizabeth, her brother, William, and her mother, Kate Lee Matthew, recently remarried (as of January 1, 1933) to Professor Ralph Minor of the University of California at Berkeley. These immediate family members were of course very close to me.

Then there were some of Dr. Matthew's siblings: his sister, "Auntie Bess" Manning, and his brothers George, Harrison, and Charles, as well as certain members of their families. There was McGowan (Mac) Matthew, the son of Charles Matthew and a descendant through his maternal ancestor of General Samuel McGowan, one of Robert E. Lee's cherished lieutenants. Mac is especially interested in family history.

Charles Olton, grandson of Robert Matthew, George Frederic Matthew's brother (and W. D. M.'s uncle), likewise is interested in family history. He furnished valuable information about his branch of the Matthew family, particularly the Dillers, and about St. Luke's Episcopal Church in Brooklyn.

Through Mr. Olton I was provided with information by the Reverend Richard F. Brewer, rector of the Church of St. Luke and St. Matthew in Brooklyn, including photocopies of church records showing the marriage of George Frederic Matthew to Katherine Diller on April 15, 1868, and the marriage of Robert Matthew to Sarah Christiana Diller on September 13, 1870.

Other sources were close friends and associates of Dr. Matthew, especially Walter Granger, Albert Thomson, Henry Fairfield Osborn, William King Gregory, and George Simpson, all of whom I knew well.

If I had had the sense to gather information for a Matthew biography in those long-vanished years, I might have garnered numerous details that could have enriched the present work. But I wasn't thinking in such terms, so various questions that have arisen during the writing of this book must remain unanswered. Even so, I learned much about Matthew over the years, particularly through my close associations with Granger, Thomson, Gregory, and Simpson.

Kate Lee Matthew Minor—my jolly, beloved mother-in-law—had many things to tell me. Bill Matthew has been able to provide information on the family home in Hastings-on-Hudson. Also I am deeply grateful to George Colbert, my cartographer son who has an encyclopedic knowledge of New York City geography, for his determination of the several William Diller Matthew residences in Manhattan.

Above all, I am indebted to my wife, Margaret, for many intimate glimpses into the life of her father and the Matthew family.

I wish also to acknowledge with gratitude the help I have received from various people (other than those already mentioned) interested in Matthew's impact on the history of vertebrate paleontology.

Dr. Randall F. Miller of the New Brunswick Museum in Saint John gave invaluable aid and information. He and librarian Carol Rosevear also provided access to many of W. D. Matthew's letters to his parents, which are catalogued in the Library and Archives of the New Brunswick Museum.

I wish to thank the archivist of the University of New Brunswick for information concerning Matthew as a student at that institution.

Charlotte Holton and Dr. Richard Tedford of the American Museum of Natural History in New York City kindly granted me access to the Matthew archives in that institution and provided copies of pertinent letters. On behalf of the museum Ms. Holton also lent me Matthew's little notebook record of his trip to Europe in 1900.

To Professor Joseph T. Gregory and to Dr. Samuel Welles of the University of California at Berkeley I am deeply grateful for information about Matthew's years there. Professor Gregory, who is working on a history of the paleontology department at Berkeley, provided some much-needed details about Bacon Hall, where Matthew was ensconced during his three years at the university. Dr. Welles was a student during some of those years and was personally acquainted with Matthew.

I thank Dr. Robert M. Hunt, Jr., of the University of Nebraska, Lincoln, for details about Agate, Nebraska, and the famous quarries there.

Professor John H. Ostrom of Yale University kindly read parts of the manuscript and encouraged me in my efforts, and for his attention I am most appreciative.

I am much indebted to Dr. Ronald Rainger of Texas Tech University for an interchange of ideas concerning Matthew and Henry Fairfield Osborn, and particularly for the pleasure of reading his penetrating study "Just Before Simpson: William Diller Matthew's Understanding of Evolution" published in 1986 in the *Proceedings of the American Philosophical Society.*

With regard to published materials about Matthew, the writings of the late Professor Charles L. Camp of the University of California at Berkeley, have been of great help. Camp was closely associated with Matthew at Berkeley.

I wish to express my appreciation to Professor Kenneth Rose of

The Johns Hopkins University for his tribute to William Diller Matthew, which constitutes the final paragraph of this book.

Finally, I must extend my appreciation and thanks to Ann Cole, who converted into legible typescript the heterogeneous manuscript pages, some handwritten, some typed, some done on a word processor, and all liberally defaced with scratchy corrections and changes.

Edwin H. Colbert
The Museum of Northern Arizona
Flagstaff

Partial Genealogy for William Diller Matthew

WILLIAM DILLER MATTHEW

Paleontologist

CHAPTER I

The Loyalists

Of those scientists who have devoted their lives to the study of fossils, William Diller Matthew is among the select few who have achieved immortality in the annals of paleontology. During most of his adult life he was on the staff of the American Museum of Natural History in New York, where he lent distinction to that world-renowned institution because of his very significant field investigations and his penetrating research studies, which had to do particularly with evolutionary development and the worldwide distribution of fossil mammals. Throughout his life he was preeminent among paleontologists the world over for his knowledge of these warm-blooded vertebrates, large and small; his mind was a storehouse containing a staggering volume of data, yet he was never so engulfed by detail that he lost sight of the grand patterns of evolutionary history. In the last years of his life he transferred his formidable paleontological talents to the University of California at Berkeley.

He was so universally recognized as one of the great paleontologists of his time that well before his fiftieth birthday his name was put forward for election to the National Academy of Science of the United States, the ultimate honor for an American scientist. But it soon became clear that Matthew could not be elected to the National Academy because he was not a citizen of the United States; he was a Canadian.

In fact, he was a loyal Canadian but never made it an issue in his life, for Matthew was essentially an apolitical person; he saw distinctions between Canada and the United States, but he viewed them with indifference. Others, however, were not indifferent to his being denied the recognition he so richly deserved because of a boundary line on a map, and so on May 8, 1916, Dr. Matthew received a handwritten letter from Arthur Smith Woodward (subsequently Sir Arthur), paleontologist at the British Museum (Natural History).

My dear Dr. Matthew:

I understand that you still remain a Canadian, notwithstanding your long residence in the U.S.A., and, if that is the case, some of us wonder whether you would appreciate the Fellowship of the Royal Society if it were conferred on you. It would, of course, involve an annual subscription of 3 pounds. If you care to be placed on the list of candidates (to be first considered in 1917), it will give me much pleasure to have the honour of taking all the necessary steps.

With kind regards,
Yours Sincerely,
A. Smith Woodward.

William Diller Matthew thus became an FRS, which undoubtedly pleased this lifelong Canadian Loyalist, although he never made a fetish of this title, as is the prevalent custom among British scientists who have received the honor.

Although Matthew was apolitical and spent his adult life "south of the border," his Canadian ties (never overtly expressed) were strong. They were rooted in a long background of Loyalist ancestors, and that background is worth exploring for a better understanding of the man.

Let us therefore go back to the middle of the eighteenth century, to George Matthew, a sturdy farmer lad living with his parents at Invergowrie in the Carse of Gowrie, near Dundee, Scotland. One day George Matthew wanted the use of a horse to ride to a fair that was being held in Dundee, but his father, also named George, as was his grandfather, said no. That was the beginning of a family saga, for young George decided that if he could not have a horse to ride to the fair, he would run away from home. Which he did.

He made his way to Broughty Ferry, the deep-water port of Dundee, where he sought out his uncle, Mr. Patullo—the brother of George's mother—a merchant and shipowner. Now that George had run away from the farm, he wanted to ship on one of his uncle's vessels. Mr. Patullo immediately got in touch with George's father, to tell him that the boy was safe in the Patullo house but wished to go to sea and live the life of a sailor. Let him go, said the father; a voyage on the ocean would cure him of all desire to live that kind of life, and he would come back to the farm.

Such was not to be the case. George shipped on one of his uncle's vessels and liked the life. Subsequently, he sailed in other ships and in time became a ship's captain.

George Matthew's horizons inevitably expanded once he left the farm and went to sea. Within a few years he married a young woman named Jane Hamlin, the daughter of a banker who lived in Cork, Ireland. By now George was engaged in the American trade, so he took his young wife to Philadelphia where the couple settled to raise a family.

Then came the American Revolutionary War, the *rebellion* to those Americans who remained loyal to the Crown. The records do not show how George and his family weathered the conflict—probably not very comfortably because they were known Loyalists. Yet they managed to survive through eight difficult years until the war ended. The Loyalists were stranded in the new nation among people who were, if not hostile to them, at least not sympathetic. By the spring of 1783 only St. Augustine, Florida, and New York City remained under royal authority, so that most of the Loyalists in the new United States were exposed to the whims and possible retributive actions of the rebels all around them.

Although the Treaty of Paris was not signed until September 3, 1783, and the British troops were not to leave New York City until late November of that year, it was quite apparent early in the spring that the occupation of New York would soon end. Consequently, between April and November of 1783 there was a massive evacuation of Loyalists to Canada, mostly from the port of New York. There were five major evacuation fleets that during those months transported some thirty thousand refugees to Nova Scotia and to the region later to become New Brunswick.

George Matthew, ship captain, participated in the "fall fleet" evacuation, carrying discouraged Loyalists to the shores of the settlement that became Saint John, New Brunswick. Many of the passengers on the transports were not overjoyed when they saw the wild, bleak coastline of their new homeland. Jane Hamlin Matthew, who arrived on a later transport, is reported to have said that this new land "should never have been taken from the bears and wolves." Another Loyalist wife, after landing at the mouth of the Saint John River, climbed to the top of Chipman's Hill to watch the departure of the fleet. She said, "Such a feeling of loneliness came over me that 'though I had not shed a tear all through the war I sat down on the damp moss with my baby on my lap and cried bitterly." [1]

In 1783 the shores of New Brunswick did indeed present a daunting prospect to those hardy Loyalists, but George and Jane Matthew never faltered. They attacked their new environment and their many new problems with the determination so characteristic of pioneers through the ages. Within a few years they had established themselves as important citizens in the small but growing community. George Matthew applied himself to the fur and lumber trade, and eventually he became harbormaster of the port of Saint John. During these years they raised a family of seven children.

One of the children was George, born on February 1, 1795, the fourth in this Matthew sequence of Georges and the first to have been born in the New World. He married Deborah Eliza Harris of Rothesay, a pleasant suburb several miles up the Kennebecasis River from Saint John. This George, like his father, became a solid citizen—a merchant and a mill owner. He and Deborah also had seven children but two died in childhood. The five survivors were, in order of birth, George Frederic, Charles, Robert, Elizabeth, and Douglas. It is George Frederic, born on August 12, 1837, who is of particular interest to us; he was the father of William Diller Matthew.

[1] Robert S. Allen, *The Loyal Americans* (Ottawa: National Museums of Canada, 1983), 62.

Little is recorded concerning the childhood of George Frederic Matthew. His formal schooling was limited to the years he spent at Saint John Grammar School; after graduation from that institution his classroom days ended. At the age of sixteen he was employed at the customs house in Saint John, where he eventually became chief clerk and surveyor, and where he remained until retirement.

If he had confined his activities to his official duties, he very probably would be long forgotten by all but a few of his devoted descendants. But this George Matthew was a most unusual man, for he elected to be, in his spare time, a geologist and paleontologist—not a hobbyist or a Sunday fossil hunter, but a serious, scholarly authority in his chosen scientific field, destined to occupy a lasting place in the history of North American paleontology.

In 1857, when George Frederic Matthew was twenty years old, he had become so dedicated to his interest in rocks and fossils that he was among a small group of enthusiasts who organized the Steinhammer Club, whose purpose was to study the local geology. The founding of the club was inspired in part by the presence of a collection of rocks and fossils assembled by Abraham Gesner, the first government geologist of New Brunswick, and contributed to the Mechanics Institute Museum in Saint John by Gesner's friends, who had purchased the collection in an effort to get him out of debt.

Sir William Dawson, one of the giants in the history of Canadian geology, had in 1855 published his great work on Acadian geology, based largely on his studies in Nova Scotia. He became interested in the Steinhammer Club, particularly in the work that was being carried on by Matthew and C. F. Hartt. They supplied Dawson with information and specimens that he used in the preparation of a revision of his Acadian monograph and in other publications on the early fossil plants of New Brunswick. At his suggestion Matthew, Hartt, and others organized in 1862 a Natural History Society in Saint John.

By this time Matthew felt sufficiently confident in his scientific abilities to commit his views to paper, and in 1863, at the age of twenty-five, he published his first piece, entitled "Observations on the Geology of St. John Co., N.B." This was a modest token of things to come, for in the course of his long life he was to publish more than two hundred scientific papers and monographs on the geology and paleontology of the Maritime Provinces of Canada—studies devoted especially to the Paleozoic rocks and fossils of New Brunswick, and to the glacial and postglacial history of that region. Many a full-time professional geologist or paleontologist would be proud of such an achievement.

One of the leading lights in the Natural History Society, he was instrumental in the establishment of the New Brunswick Museum, today housed in an imposing building on Douglas Avenue in Saint John. His position as one of the founding fathers of both the society and the museum is still well remembered; many specimens that he collected are part of the holdings of the museum, and for several months in 1987 a special exhibit commemorating his life in science was on display in the large entrance hall of the building.

These pioneering studies in his chosen field of science eventually received wide acclaim; he was awarded an M.A. and an LL.D. from the University of New Brunswick and an honorary doctorate from Laval University. He was a charter member of the Royal Society of Canada and for many years a Fellow of the Royal Geographical Society of London, which awarded him the coveted Murchison Medal. He was well known in geological circles, not only in North America but also in Europe and other parts of the world. Needless to say, he carried on an extensive correspondence with scholars in many lands, often contributing advice and opinions to his correspondents.

One little-known but rather amazing fact of his professional life is that on his own he became fluent in the Swedish language—so fluent that at the request of the United States Geological Survey he provided that organization with translations of Swedish geological and paleontological papers.

One cannot help but be astounded by the prolific scientific career of George Frederic Matthew, accomplished as it was in his spare time, since he never relinquished his full-time position at the custom house and cheerfully gave attention to his large family and to his home as well. Among other things he maintained a productive vegetable garden each year, almost a necessity for him with so many mouths to feed.

On April 1, 1868, George Frederic Matthew had married Katherine Diller of Brooklyn, New York. This leads to a side trip in the story of William Diller Matthew and his forebears. Who were the

Dillers, and how was it that George Frederic Matthew married a young lady named Diller, who lived five hundred miles from Saint John?

Let us go back to Caspar Diller, born between 1670 and 1675 in Alsace, of protestant parents. When he was still a young boy he was taken by his father to Holland, where the family had fled to escape religious persecution. Caspar later went to England, where he married. He took his English wife to Germany and they settled near Heidelberg. In 1729 he emigrated with his family to America, to become established in Lancaster County, Pennsylvania, as did many German immigrants in those days. He was the father of seven sons and three daughters.

One of the sons, Philip Adam Diller, married Elizabeth Ellmaker of Earl, Pennsylvania, and from this union eight children were born. This brings us to Leonard Diller, one of the eight, who married Mary Magdalina Hinkle of Hinklestown, Pennsylvania (one may assume that her family was descended from the founder of that community), and these Dillers produced six children. One son, George, married Lydia Souder, and one of *their* eight children (the Dillers seemed to run to eight children or thereabouts) was Jacob William Diller.

Jacob Diller, educated for the ministry, was for many years rector of St. Luke's Episcopal church in Brooklyn. He married Angelina Van Nostrand of Brooklyn, and they followed the Diller pattern by having seven children. Among the seven were two daughters, Katherine Mary and Sarah Christiana, and they are of especial interest to this story because they married two Matthew brothers—Katherine wed George Frederic and Christiana (as she was known in the family) married Robert. How did it happen that two brothers in New Brunswick should marry two sisters in Brooklyn? No clues exist in the papers of George Frederic and Robert Matthew, and contemporaries have long since departed. Consequently, some speculations are in order.

George and Robert had a cousin in Saint John named James Fowler. In spite of spending his early years in high latitudes amid other Fowlers and Matthews, James developed an interest in Cuba, where he bought some property and became involved in raising sugar cane. Robert, according to a letter written by George Frederic in 1918, practiced law as a young man in Saint John and then went to New York and "engaged in mercantile business." James Fowler may have invited Robert to join him in the Cuban venture; certainly Robert spent much of his later adult life as a sugar cane planter on that Caribbean island. However, it seems likely that when he first went to New York Robert lived in Brooklyn, a pleasant residential borough, and there he must have met Christiana Diller, perhaps through attendance at St. Luke's Episcopal Church. That was the beginning of a chain of events that, among other things, resulted in the marriage of George Frederic Matthew and Katherine Diller in 1868.

Eight children were born to George Frederic and Katherine Diller Matthew; in order of birth they were Eliza Katherine, William Diller, (born on February 19, 1871), Elizabeth Mary, George, Harrison Tilley, Robert Theodore, Charles Frederic, and John Douglas. All of these children except Eliza—who never married—and Robert—who served as a lieutenant in the 60th Battalion, Victoria Rifles of Canada during the First World War and was killed in action at Ypres in 1916—emigrated to the United States to make their individual ways in life. Things were just too difficult for them in New Brunswick. But before their separate emigrations southward they enjoyed impressionable years together as members of a large, close-knit, loving family.

For them it was a frugal life, because G. F. Matthew's salary as an official of the Canadian government was not munificent, and there were numerous mouths to be fed and bodies to be clothed. Consequently, at early ages the various Matthew children learned the necessities and virtues of family cooperation—as they had to if they were to survive.

There is a letter written by William when he was a small boy to his younger sister, Elizabeth, asking if she would rather have an orange or ten cents for her birthday present. Even on this tiny scale the letter is illustrative of the feelings and actions of the Matthew children—feelings and actions that continued in their adult lives. Through the years the Matthew siblings were always very supportive of each other; their Scottish heritage and their immediate circumstances taught them that despite having to live frugally, there was, nonetheless, not just a duty but also a real joy in giving as generously as possible to one another.

Such was the background of William Diller Matthew, a North American paleontologist of great distinction, whose achievements are hailed on both sides of the border. He was a modern Loyalist—half Canadian, half American—with a broad, pragmatic, and tolerant view of life in an interdependent world. It is to this life that we now turn our attention.

CHAPTER 2

A Maritime Boyhood

*T*he birth of William Diller Matthew on February 19, 1871, was part of a pattern—a pattern that seemed designed to involve the appearance of a new child every other year. George Frederic Matthew and Katherine Diller had been married in 1868, and even though little William appeared on the scene only three years later, he had been preceded by his then two-year-old sister, Elsie. After William's arrival came the sequence of Matthew births that resulted in a total of eight siblings.

Intertwined were the pattern of frequent births, the demands of governmental and scientific activities, the necessity of providing food, clothing and shelter for more and more family members, and the inevitable, nagging problem of finances. George Matthew's salary at the custom house was modest, yet it had to suffice. So from the very beginning there was a necessity in that household for making every penny count. Economy was always uppermost in Matthew thinking and planning. Perhaps the Matthew family was not poor, but the life-style in that household was necessarily rather frugal.

Nevertheless, the Matthews lived a good life despite circumstances that seem from this distant view to have been unfavorable to such a life. They did not think of themselves as being disadvantaged in any way; they quite realistically saw the world around them as hard, perhaps even harsh at times, so they governed their lives accordingly.

The environment itself was hard. Saint John, New Brunswick, in the latter part of the nineteenth century was a small city of limited resources. The basis for its existence was for the most part lumbering, shipping and shipbuilding, some manufacturing, and fishing. Although agriculture had its place (after all, the first settlers had to scratch a living out of the soil), there was not a large agricultural hinterland beyond the city. The wild and rocky shore that had initially so discouraged those hardy Loyalists who had fled to the north from the nascent United States was still rocky, if somewhat less wild. It was not a countryside upon which to grow crops on a large scale.

The setting was and is magnificent, with a broad, picturesque estuary—the confluence of the Saint John and Kennebecasis rivers—surrounding the city, with steep, wooded hills as a backdrop, and with the dramatic flow of the tides of the Bay of Fundy rolling in and out. Yet a close look at nineteenth-century Saint John reveals a city that architecturally was far from picturesque. All along the hilly streets were wooden buildings, square, drab boxes that had an appearance of insubstantiality. There were some stone and brick structures, but such buildings, largely in the central business district, seemed overwhelmed by the crowded frame houses that extended in serried rows along the city streets.

Yet in spite of its unprepossessing appearance, Saint John in the late nineteenth century was a vigorous town. It was not large at that time—its population was about forty thousand—but its influence was widely felt. Its ships were known throughout the world; it was an important port. Furthermore, the time was one of growth and development, a period that is recalled as a "golden age" in the history of Saint John. Then, toward the end of the seventies there were massive reversals in the fortunes of the city, due in part to a world-wide depression, and in part to the great fire of 1877, which marked the end of that first golden age of Saint John.

Almost every city, great and small, has had in the course of its history at least one great fire. The Saint John fire of 1877, while not as renowned as the London fire of 1666, or the New York fire of 1835, was nonetheless an overwhelming catastrophe for the small city on the Bay of Fundy. Even today, more than a century after the event, the story of Saint John is commonly told as a two-part tale: the city before the fire, and the new city that rose from the ashes of its predecessor.

The fire erupted on June 20, 1877—"Black Wednesday" in the history of the city. It was a beautiful but windy day. About two o'clock in the afternoon sparks from an active lumber mill located on the waterfront ignited some hay stored nearby, and almost immediately the fire was of uncontrollable size. In the words of one witness, "When the fire took hold of the hay the flames went like a flash and the building became all at once ... one mass of flames." The fire spread across the city with explosive force, so that block after block of wooden warehouses, houses, and outbuildings were consumed. In the end some twenty thousand Saint John citizens were rendered homeless, and the financial losses were in the millions of dollars.

But as so often happens in the case of metropolitan fires, the city experienced a rebirth in which much insubstantial wood construction was replaced by stone and brick, although, to be sure, there were also new wooden buildings erected to replace those that had burned. Nevertheless, Saint John after 1877 was a more substantial city than it had been in earlier days.

The fire wiped out the George F. Matthew home and the custom house; particularly saddening, it also destroyed George Matthew's library, his paleontological collections, and his notebooks and manuscripts. The furniture and other household goods that went up in flames could be replaced, but the specimens and the books, notes, and manuscripts, were gone forever. The pioneer paleontologist of New Brunswick had to begin his scientific life again.

This happened when William Diller Matthew—Will or Billy to his family and close friends—was six years old. The traumatic experience watching their home and its contents devoured by the flames was a bitter experience for George and Katherine Matthew, and it surely had some effect on the children, Elsie, Will, Bess, and young George, although for the younger two the holocaust probably was dimly understood. For Will it was real enough; he was heard to remark (with an air of sagacity surprising in a six-year-old) that the fire reminded him of what he had learned about the London fire.

Only a year before the fire Will had contracted scarlet fever, which in those days was a much more serious disease than today, and he was, in the parlance of those times, a "sickly" child, so perhaps the forced displacement of his family was felt by him more acutely than by his siblings. Whatever the reason, a decision was made to send him away for a year, possibly to benefit from a tranquil environment while the rest of the family went through the stress and the hurly-burly of getting resettled in a new home. Furthermore, another baby was expected soon, which of course would add to the complications of family life. And so in September of 1877 young Will Matthew was taken to Brooklyn, New York, to spend the coming fall, winter, and spring with Diller relatives.

Those Diller relatives were members of the family of the Reverend Jacob William Diller who it may be recalled, had married Angelina van Nostrand of Brooklyn. Soon after his marriage they had moved to Middlebury, Vermont, where for four years he was rector of St. Stephen's Church, and where his two eldest children, Matthew Augustus and Elizabeth Lydia, were born. The family then moved back to Brooklyn, where Jacob was established as the rector of St. Luke's Episcopal Church. It was here that the younger Dillers were born— Katherine Mary, who later married George Frederic Matthew, Sarah Christiana, who later became the wife of Robert Matthew, and a still younger daughter, Ellen Caroline.

In 1877 Jacob William Diller was a widower, but it would seem that his son, William, and family were living at the St. Luke's rectory with the widower. Thus little Will Matthew was to spend his year away from home with a grandfather, uncle and aunt, and various cousins. His first letters to his mother were written from the rectory.

Brooklyn, N.Y.   (September, 1877)

dear mama. I had a very pleasant time on my journey. And when I got home I found this silver. which is for Elsie. with my love. and I found some shells only I lost them. dear mama this is all I can think of now.

your loveng son willie.

Soon afterward in this same month there was another letter.

St Lukes Rectory

My dear mama

I go to school morning and Afternoon. and we went to coney island and had a feine time. and I have three little papers called the cristan soldeir.

your loving son willie

Not bad for a six-year-old boy just starting his formal schooling. But his aunt Lydia (or Lilly) who was then also at the rectory, was not entirely satisfied:

Dear Kate—

This was written on Tuesday, but I waited to add a line or two. This is a very badly written letter of Willie's, but it was the best I could get out of him. The fact is that when he comes in from school at three o'clock he wants to play and I think he ought to, so I don't like to insist upon his writing. Hereafter I think Saturday morning would be the best time. He still finds many new and curious things here and his head is almost turned with the variety. He is on the whole very much improved by going to school, and has gotten quite into the spirit of being in time and not losing credits. He has taken a good place in spite of his slowness in arithmetic, and is in class with much older boys. I am thus far quite satisfied with the experiment. We are going to keep him here for a while until Molly is quite strong, and Chrissie and Rob [Will's aunt and uncle, already mentioned] arrive and then we can let him spend the rest of his time with Molly. Father is perfectly delighted with his reading and thinks there is something wonderful about him.

Your loving sister
Lilly

Among the Diller cousins the two sons of William and Mary, namely Francis (Frankie) and Alfred (Alfie), in short order became playmates and boon companions of the little newcomer from Canada. The three boys were constantly busy with projects of their own devising, so that the hours after school and on weekends passed quickly, and Will would seem to have been spared the trauma of acute homesickness. He managed, however, to write letters to his family far to the north, building anew after the great fire. Perhaps he squeezed the composition of these missives in between his busy activities with Frankie and Alfie; perhaps the squeezing resulted from gentle pressures administered by Aunt Lilly. It is interesting to see that these later letters, written after some months in the Diller household—when he had become fully adjusted to his new environment—were longer and more informative than his first attempts.

Dec 6th 77

Dear Mama

How is Elsie? I would like to know too what things she has to play with? how is she getting on with her lessons? I hope she is getting on very well is she?

Frank is going to set up a telephone between this house and another boys house. [Frank must have been *au fait* with Victorian technology. Bell and Watson had made their first basic discovery of the principle of telephonic communication on June 2, 1875, and the public was introduced to the telephone only the next year, at the Centennial Exposition in Philadelphia. One wonders what kind of a system Frank was going to rig up between the two houses.] And how are Bessie and George getting on? are they getting on very well? I and Alfy as soon as I have finished this letter will play carpenter.

And just before Christmas Will wrote to his mother, extending "love and kisses" to the family, and informing his mother that "I send this little present to you which I made myself." On this frugal, loving note the old year drew to a tranquil end for Will Matthew, far removed from the turmoil of rebuilding and resettling in Saint John.

With the coming of the New Year, and the Christmas holidays behind him, Will had to be reminded that it was time for another letter. But the muse was not with him.

Feb. 8, 78

Dear Mama

I hope you are well. I do [not] know of anything to write. good by from your loving son

Willie

Once again an aunt—this time his aunt Mary—felt that steps had to be taken.

Feb. 9th/78

Dear Katie—

After an hour's sitting, the foregoing is all Willie accomplished, and as he wasted so much time I thought I'd shame him of daw-dling—by sending it to you. I have made *some* improvement in this direction—I wish you would think it best to let him know I sent it, and lead him a little lesson on it. It will do much good from you, since you are so far away—Baby is so lovely. It seems as if she could hardly belong to us,

Affectionately your
Sister Mary

("Baby" was none other than Angela Diller, who in years to come became widely known and respected in musical circles as a teacher. She was a founder of the Diller-Quale School of Music in New York.)

Brooklyn by now was familiar home turf for Will Matthew, and with various Diller and Matthew cousins at hand—Christina and Dorothy, the daughters of Robert and Christiana Matthew, were at this time in Brooklyn—there was ample company to fill the non-school hours. Will writes to his mother about playing with Frank and Alfie Diller, and Christina and Dotty Matthew. He obviously was not the least bit lonely.

Spring came, and the tree-lined streets for which Brooklyn in those days was famous were avenues of dappled shade, and Will Matthew's first year of schooling was drawing to a close. Not long before he was to return to Saint John, a gathering of the Brooklyn families occurred. The occasion was "Decoration Day" (Memorial Day, now). Will tells about it in a letter dated June 4, 1878.

We had a fine time on Decoration day . . . all but a few of us played croquet . . . All of us played blind mans buff except Alfy and I. Uncle Rob had a box what would make sounds of music and tickets were made each of us had to show a ticket before we could come in and a little while after that we had a show of a magic lantern. . . . We are having strawberries now.

Good by from your loving son Willie Matthew
P.S. give my love to Papa and Elsie and George and Harry.

(Harry was the new baby, Harrison Tilley Matthew.)

During those months when Will was experiencing his first year of school in Brooklyn, the Matthew family in Saint John was setting things to right in a city that looked like a war zone. Through at least a part of that winter of 1877–78 they sought refuge at Ashbrook, a summer place (belonging to brother Robert) on the shore of the Kennebecasis River, fifteen miles or so north of Saint John. It must have been rugged at best for all involved. A family legend (probably true) has it that George Frederic Matthew each morning made a cross-country trip of a mile or so to the railroad station at Quispamsis, to take the train to town to his work. And of course each evening he made the return journey, arriving on skis at his front door.

But such a picturesque style of life did not last for long; the family in due course became established in a rather commodious house known as "Hillside," located as the name indicates on the side of a hill in a section of Saint John. Here Will Matthew lived for a decade after returning from Brooklyn. (The site of the house is still to be seen—a wooded area on the hill, the house having burned down some years ago.)

For a greater part of the next decade Will continued his education in the local schools. He was a precocious scholar, so much so that he graduated from high school at thirteen.

Contributing further to the impression he gave of studiousness was the fact that even at this early age he needed spectacles—perhaps in part as a result of reading in poor light. It would seem, however, that his eyes were naturally weak; throughout the remainder of his life he was virtually blind without his glasses.

He was too young, his parents thought, to enter the University of New Brunswick in Fredericton, up the river from Saint John. Therefore he was apprenticed to work in a Saint John law office for three years, to fill the time until, at the still tender age of sixteen, he was adjudged sufficiently mature to begin his higher education.

In the meantime he had been receiving a most valuable, albeit informal, course of instruction in earth history from his father. It had begun when he was a young boy; in 1877, when he was six but before his Brooklyn sojourn, Will had the distinction of discovering the largest of all trilobites. These ancient fossils—the remains of hard-shelled arthropods related in a general way to the modern king crab (or limulids) and distinguished by their three-lobed bodies, transversely segmented, and a large head region (or cephalon) bearing a pair of prominent eyes—are among the most ancient of fossils. They are especially typical of Cambrian sediments, which date back almost 600 million years.

It was in Cambrian shales, right in the middle of Saint John, that little Will Matthew discovered this giant trilobite, some sixteen inches in length and twelve inches in breadth. His father helped him dig it out of the shale, and in 1888 George Frederic Matthew described it in the *Transactions of the Royal Society of Canada,* naming it *Paradoxides regina.* As the elder Matthew wrote in his description, "This trilobite is honored with Her Majesty's title, as that of a sovereign, who, during the many years of her reign, has greatly fostered science and art."

In the type description it was stated that *Paradoxides regina* was "found by W. D. Matthew," with no indication that at the time the discoverer was a boy.

The type specimen of *Paradoxides regina* was W. D. Matthew's first significant fossil find.

Shortly after *Paradoxides regina* had been removed from the Cambrian shales in which it had rested for so many millions of years, and some eleven years before it was formally named and described, came the great fire. The fossil was partially destroyed in the fire, but fortunately most of the specimen was unharmed. Today it is one of the treasures of the New Brunswick Museum.

Undaunted by the great fire, George Frederic Matthew had resumed his studies in geology and paleontology, and had turned his attention toward expanding his fossil collections. Will was a junior partner in some of these efforts, even before his graduation from high school. He not only accompanied his father on fossil hunting trips but also made collections on his own initiative. For example, he would carefully search discarded ballast from the many ships that came into Saint John, thereby acquiring fossils from distant provenances. "I have been arranging my shells now and then. I have quite a number of shells which came from Buenos Ayres(?). I got them in ballast" (letter to his mother dated April 26, 1885).

At fifteen Will was working at a lawyer's office in Saint John, but he was also helping to manage the house and a part of the family while his mother, his sister Bess, and his brother Robin, were away on a visit "at Uncle Rob's." Whether Kate Matthew and her two children were visiting Robert and Christiana in Cuba or in Brooklyn is not evident from the correspondence. In either case, Will and his father remained behind to cope with the Hillside house as well as with the younger boys, minus Robin, namely George, Harrison (born in 1878, when Will was living in Brooklyn), and Charlie. Charlie was presently the baby of the family, but he was soon to relinquish that role to Jack, the last of the Matthew children, in November of 1885.

A certain Julia, evidently a maid or a household helper of some kind, assisted in keeping this lively family on an even keel. Even so, Will, as the eldest son, felt that many responsibilities rested on his youthful shoulders.

March 5, 1885

Dear Mother

By the time you get this letter I suppose that you and Bessie and Robin will be comfortably settled at Uncle Rob's. You would not imagine how we miss you. The house seems very empty without you. Charles misses Robin more than anyone else, I think. He walks around the house and seems, very lonely. Often when he sees me he asks me to come and play with him. I do not wonder at his feeling so for when the boys [George and Harrison] are away at school he has no one to play with except Julia and myself and neither of us has much time to spare. However he is very good.

I counted my specimens a couple of days ago and found out that I have over 500 specimens. I have about 275 minerals, 100 shells and 128 fossils. I have also about 20 curiosities etc. This is just a hasty count and when I have a Catalogue, that I am making, finished I will send you a more exact account of them.

I remain
your affectionate son
Willie

Here one sees the young naturalist revealing an aspect of his nature that was to loom large in his professional life as a paleontologist. Matthew was, during his long paleontological career, a careful and conscientious cataloguer. He catalogued and labeled thousands of fossils, particularly of mammals, and in so doing became intimately acquainted with the anatomical details of those specimens. This was in part the basis of his encyclopedic knowledge of fossil vertebrates.

Although Will Matthew in 1885 had much satisfaction with his collection, he was at times less than pleased with the doings of his brother George. In two letters to his mother there were some complaints.

April 19, 1885

Master George made another attempt to set the house on fire yesterday. He found a lot of paper trash on the coal scuttle which he set on fire, and then retreated, and shut the door. Julia found it out by the smell. Of course George got no punishment.

April 26, 1885

George owes Papa 6 cents and has owed it for about 3 months, for, though he has plenty of chances, he says he 'don't want to,' otherwise he is too lazy. A couple of weeks ago he was at my watch and he got water into the works and made the hair spring so rusty that the watch cannot be fixed. It would be very tedious if I were to try to give an account of even his principal meddlings.

It *was* exasperating, yet about twelve years later, when a student at Columbia College, George was given financial help and guidance by his loving elder brother.

On April 26, 1885, Will wrote to his mother that he was "going to work next Friday with Mr. Lee. I saw him a few days ago and he said that I could come down about the middle of the morning and he would see that the work was not *too hard* for me—So I will not be very badly used, I think."

And in a letter dated May 5, he described some of his duties.

The work is not very killing, in fact it is the next thing to nothing at all. My principal work is to "keep shop" for him. Otherwise I stay in when he goes out and tell callers where he is and when he will be back and ask them to sit down and entertain them. The last is the worst part of the business. [Will Matthew was never adept at social chit-chat, so the task of entertaining strangers must have been a real chore for him.] During my spare times I read Gibbon's "decline and Fall of the Roman Empire" and try to struggle through "Cornelius Nepos" without a *dictionary*.

Not long after receiving this letter Kate Matthew returned home, bringing Bess and Robin with her. Those spring, summer and fall months were especially busy, with the family, reunited and a new member expected late in the year. The last of these Matthews, John Douglas (or Jack, as mentioned), was born on November 8, 1885.

For Will the remaining months of 1885 and 1886 were a time of waiting until he was old enough to enter the University. Yet this period in his life was anything but static, in part because he was living in a busy household, and in part because—being a Matthew—he was constitutionally inclined to fill his days with useful thoughts and useful deeds, such as reading Gibbon.

CHAPTER 3

Fredericton Years

*W*ill Matthew spent the summer of 1887 in Saint John—at home, where the eighth Matthew sibling, "Baby Jack" added to the complexity of life in an already busy household, and in town, where he continued his duties at the law offices of Mr. Lee. Then October loomed, and it was time for his first year at the University of New Brunswick.

Something should be said about the university and its history and institutional nature in 1887. In the middle of the nineteenth century the few Canadian universities then in existence could trace their beginnings back to either English or Scottish origins. In the schools of English derivation the sciences were offered largely as cultural subjects, part of a curriculum based on the classics, the humanities, and mathematics. The teaching of science by disciplined, compulsory lectures and rigorous laboratory exercises, as developed in the German universities, had not as yet been adopted. So it was that the College of New Brunswick, founded early in the nineteenth century, had a tradition of classical education with mathematics, but no science. In 1828 the school was renamed King's College. Its president at that time was Dr. Edwin Jacob, an Anglican divine and a graduate of Oxford, who saw as his purpose the development of this backwoods college in the classical English mold. But such a school was hardly suitable in the pioneer environment that prevailed in nineteenth century New Brunswick.

Indeed, Sir Charles Lyell, the great British geologist (who influenced important aspects of Charles Darwin's thinking) made some comments about King's College, which he visited during his midcentury tour of North America. King's College, he said, was "rendered useless and almost without scholars, owing to the old fashioned Oxonian of Corpus Christi, Oxford, having been made head, and determining that lectures in Aristotle are all that the youth in a new colony ought to study, or other subjects on the strict plan which may get honours at Oxford" (letter from Lyell to Leonard Hoerner, September 12, 1852).

Then three young Scots joined the faculty of the college in Fredericton—David Gray and James Robb from Edinburgh in 1837, and William Brydone Jack from Saint Andrews in 1840—and the result was a beginning of a science curriculum at King's College. In time William Jack became the president of the college, after which, owing to measures taken by the lieutenant governor and the provincial legislature, the institution in 1859 was reconstituted as the University of New Brunswick.

In 1861 a new science professor joined the faculty of the new university, an American named Loring W. Bailey. He had studied at Harvard under Louis Agassiz, Asa Gray, and Josiah Parsons Cooke and thus had a broad training in the natural sciences, but his primary interest was in geology. It was his good fortune to soon become involved with George Frederic Matthew and C. F. Hartt on collaborative studies of New Brunswick geology, and he proved to be a dedicated and prolific author of geological papers.

Thus in the autumn of 1887, when Will Matthew journeyed to Fredericton, he was well known to Bailey, his father's friend and collaborator. In fact, plans had been made for Will to live in the Bailey home. He made the trip by boat, up the Saint John River to Fredericton.

Oct. 2, 1887

My dear Mother

As you will see by the heading of this letter, I arrived here all right. I was not so lonely on the trip as I had expected, as there were my two companions from St. John, Percy Hanington and Will Vanwart and Miss Henry of St. Stephen who is going to take the course. The fog was pretty thick at first but after we passed Oak Point it cleared away nearly altogether and I could see the river and surroundings very well.

Prof. Bailey and Margaret were at the pier and had a hack ready for me, for which I paid 25¢.

I am not yet installed in my own room, but have one just next to it and of the same shape and size. It is a very nice room, large, comfortable, and quite well lighted. I have a good view over the garden and fields to the College.

They were all very kind, especially the Mrs. and told me to make myself entirely at home; which I accordingly did.

Will, a slightly built young man of fair complexion, with rosy cheeks, sandy hair, and brown eyes behind steel-rimmed spectacles, was required to take entrance examinations before he was formally admitted to the university. He did so, passed, and began his classes in early October. The curriculum embraced English, French, German, the classics, mathematics, and science (whatever that might have been). Evidently he was adept in his scientific studies, for in a letter written home on December 13 he said that his science notebook was in great demand among the boys in his class; he went on to say that he could not lend it indiscriminately to all who wished to benefit from the notes he had made. (It is obvious that he was a superior scholar; moreover, his notebook must have been desired in part because of his beautifully clear handwriting. From his boyhood until his death everything he wrote was remarkably legible. He never used a typewriter, nor did he need to.)

By the time of the December letter he had, to his satisfaction, settled into the routine of being a student at the university. But two extracurricular problems bothered him, one being the constant worry about money, and the other being the small annoyances of life in the Bailey household.

As for money—a perennial problem for students—it is revealing, and pitiable in a way, to see how parsimonious he was forced to be. Of course, a century ago a dollar went a long way; even so Will had to count pennies with the greatest of care. This was a problem he had during his whole time in Fredericton, a problem he solved month by month with economies of the strictest nature. For example:

Since I came here I have been trying to keep my outside expenses at the lowest limit I can. But the books I have to get, and if I did not pay my share of the class expense I would fall into disgrace. About 20 or 25¢ which I have spent treating the boys at different times have been the only unnecessary expenses. "When in Rome do as the Romans." Shoestrings, collar buttons, stationary, etc., etc. I include as necessities, though they do not amount to very much in proportion to the books and class expenses.

(letter to his father, October 6, 1877)

He could handle his money problems on his own, but the problems of life among the Baileys required tact and patience that at times he found almost beyond his abilities. Soon after having become settled at the Bailey home, he had to let off steam in a letter to his sister Bess.

The small Baileys are most abominably spoilt, so indeed are the others. The "cheek" they give their Pa and Ma is terrific. I shouldn't be surprised if the word "please" came into conversation, to hear one or both of them ask what it meant. Margaret is not a bad girl on the whole, especially when she tries to be pleasant. Loring is good enough in a way, but takes pride in his "fast" habits, goes out (nobody knows where) every night. I do not like him much.

The professor is very pleasant, quiet, hardly says a word, and is what you might call easy going. Mrs. Bailey is also pleasant, but scolds the children instead of spanking them. There now, I have vented my ill humour on the inoffending (to me) Baileys, for they have been as nice as possible so far as I am concerned.

(letter of October 17, 1887)

Two days later in a latter to Bess he fulminated some more, and at considerable length. His strictures will not be repeated here; suffice it to say that he could hardly tolerate Margaret's laziness, Loring's unsocial behavior ("never at home except to his meals"), and the crying and frequent whining of the two little children.

One should remember that Will Matthew was not expressing the sour views of a solitary bachelor accustomed to peace and quiet, for he had lived with six younger siblings and had ample experience of life in a house filled with lively, noisy children. Consequently, in spite of his privately expressed misgivings he was able to put up with the Bailey ménage, trying though it must have been at times.

Mrs. Bailey's propensity for having parties was for him a cross to bear. These parties cut into his time and his peace of mind: "Mrs. Bailey is going to have a party on Thursday. I am awaiting the event in fear and trembling, for I know I shall make any number of blunders" (letter to his mother, January 10, 1888).

Yet one should not give too much weight to his complaints about life in the Bailey household, despite the occasional strongly worded outburst. Will Matthew was fond of the professor and his wife, and he truly appreciated their assistance and their many kindnesses, even including the invitations to her parties—gestures extended to him out of sincere goodwill.

When winter came on Will indulged in various outdoor activities characteristic of that refrigerated northeastern corner of North America. Winter in Fredericton was not so much to his liking as winter in Saint John, where there were hills for sledding, and where at night one could skate mile after mile on the frozen Kennebecasis over ice so black that the skater, looking down, could see nothing beneath his feet but starlike crystallizations that glittered from below the invisible surface of the ice. And overhead, in the cold sky, the northern stars would sparkle across the arch of the heavens. Will and his sister Bess and other Saint John friends would sometimes on such a night skate from the city up the estuary toward Ashbrook—an adventure long to be remembered. Such recollections may have been in his mind as he sampled the outdoor winter pleasures of Fredericton.

One reason he found winter in Fredericton less than perfect was the extreme cold typical of that city's inland location. On January 27, 1888, he reported that "we have snow enough now—about two or three feet of it—and cold enough too, for it was 22 degrees below zero at 10 o'clock this morning by a thermometer with the hot sun shining on it. It is so cold and dry that one fairly gasps for breath if one tries to speak or open the mouth" (letter to his mother, January 27, 1888).

Yet he did skate and sledded at times, even though exposure to the frigid air played havoc with his tender skin. (Matthew's sensitive skin was a problem with which he had to contend throughout his life. During his years of paleontological explorations in the American West he suffered, year after year, from sunburn. He never tanned but rather burned and blistered, so much so that, despite protective clothing and ointments, his months in the field were periods of epidermal torture.)

Even though he was very busy at the university with his studies and engaged in various extracurricular activities, the long, cold winter got on his nerves. His elder sister, Elsie, came up from Saint John for a visit in late January, yet her presence could not entirely allay a feeling of depression that was very real—as shown in his letters home.

> I am suffering here for some variety. There is nothing to do but read—or go to the rinks, toboggan slides, or coasting or walking. Reading I am tired of, toboggan slides cost too much—rinks I do not care for and I do not care very much for coasting except when I am on my own sled—and walking I am also tired of, for there is nowhere to walk.
>
> Generally speaking I am sick of this winter weather—out of doors bitterly cold so that one has to wrap up thickly like a mummy and is never comfortable, and indoors, dry and hot and close and generally uncomfortable. Now all things except the people are dead or hidden away. I often wish I was like a bear (parentheses by the reader—'so he is') and could go to sleep in the autumn and never wake up till spring.
>
> If I said I suffered from want of variety I meant to say agreeable variety. As it is I have a certain variety in sewing up moccasins, sewing on buttons, etc., but that is not exactly the variety to which I referred. I have no friends here—I have at most only two or three—and it is winter.
>
> I have not been to a party yet, except that one that Elsie snared me into at the Tabor's, and I do not intend to go. Why should I when if I do go I cannot dance, or play games, and can only stand around and make myself and others uncomfortable.
>
> (letter to his mother, January 27, 1888)

It was the winter of his discontent.

One wonders if these were the remarks of a young man whose intellectual potential was not being sufficiently challenged. Or perhaps it was only the miserable winter weather after all.

With the coming of March there were plans for the next academic year and a consequent lifting of the spirits. In a letter to his mother Will wrote:

You have heard, no doubt from Father about the arrangements concerning my staying here and trying to graduate. It will give me plenty of work, but I think I can manage to pass if I work hard. As a student of the second year is allowed to omit one pass subject, always with the approval of the faculty, I might omit Latin, I think. It would, perhaps be better to leave out that than Logic, for this latter is certainly of some use to any man, even in a science course.

(letter to his mother, March 14, 1888)

He did stay on at the university for another year, to graduate and receive a B.A. degree in 1889. This date is accurate but requires an explanation. How did Will Matthew complete his undergraduate education in two years?

During the middle years of the nineteenth century King's College—which in 1859 became the University of New Brunswick—had a three-year curriculum, as was prevalent among English universities. In 1885, with the retirement of Professor William Jack as president and the appointment of Thomas Harrison as his successor, the university curriculum was altered from three to four years, thus coordinating its program with those of most North American universities and colleges. Even so, how did Will Matthew, who entered the University after this reform had been instituted, complete his degree requirements in half the allotted time?

Thanks to some help from Dr. Randall F. Miller of the New Brunswick Museum, the matter has been clarified. According to Dr. Miller, who at my behest interviewed the archivist at the Harriet Irving Library, University of New Brunswick.

W. D. Matthew *did* finish his degree in two years. I spoke with the library archivist, and she told me that it would not have been unusual for someone to have completed their university courses in two years. . . . She thought it was probable that a student could have written entrance exams to have given him advanced standing. She said there were students known as "freshie-soph" who, I assume, were combining their first and second years into one. . . . I suspect [Matthew's] background and L. W. Bailey's knowledge of his early education helped him gain advanced standing.

(letter to the author, May 18, 1989)

Earlier it was mentioned that Matthew was required to take some examinations in the fall of 1887 before being admitted to the university, and as a result of his outstanding record at these examinations it may be presumed that he became a "freshie-soph" during his first year in Fredericton. Likewise, as indicated in his letter of March 14, 1888, to his mother, he may have taken further examinations, and omitted some subjects "with the approval of the faculty" and thereby completed the second half of his college curriculum in his second year. Whatever the details, we know that in the spring of 1889, just a few months after his eighteenth birthday, he was a very young Bachelor of Arts.

One might suppose that having had so brief an undergraduate career he would have been ill-prepared for higher academic studies. Perhaps in some respects this may have been so, but in many ways Will Matthew at the age of eighteen was an unusually well-educated young man. He was an exceptionally studious, and, as we know in hindsight, a brilliant person who acquired much of his learning outside college walls. Furthermore, he had attended a small institution of higher learning—perhaps more like a liberal arts college than a full university—where he had the benefit of close contacts with his professors. These teachers could, in such an intimate setting, more readily pass along their knowledge and firsthand experiences to students.

It can be argued that standards of collegiate education then were not as rigorous as today, but that is highly debatable. Of course the level of knowledge was not then so advanced as it is now, especially in the sciences, but the mental disciplines were just as stringent. Indeed, as has been stated by Richard A. Jarrell in his penetrating essay entitled "Science at the University of New Brunswick in the Nineteenth Century" (a source of some of the remarks here), "One might note . . . that the Victorian textbook was usually far more difficult and sophisticated than those of today."[1]

In the spring of 1889 Will Matthew looked ahead with confidence to the pursuit of study at the graduate level. He had a background in

[1] Richard A. Jarrell, "Science Education at the University of New Brunswick in the Nineteenth Century," *Acadiensis: Journal of the History of the Atlantic Region* (Department of History of the University of New Brunswick), vol. 2, no. 2 (1973): 70.

the classics superior to many modern-day students, he had a solid grounding in the basic sciences, and he had advanced, sophisticated knowledge, especially for so young a person, of geology and paleontology—the result in part of years of association with his father, George Frederic Matthew.

He was ready to delve deep into the natural history of the earth, in a literal as well as in a figurative sense, for he had decided to become a mining geologist. For this purpose he turned his eyes toward the Columbia School of Mines in the City of New York.

As a postscript to his Fredericton years it is revealing to read the comments published about him, as a member of the graduating class of 1889, in the *University Monthly* of the University of New Brunswick.

> Next come W. D. Matthew the genius of '89. He is a native of St. John, N. B. and is a son of Mr. Matthew of the Customs Service. He has lived in an atmosphere of Science all his life, and knew more about that study on entering, than most students do on leaving college. To tell the truth he was well up in all branches. 'Will' is one of those fellows who never make any fuss or show, but go about their work quietly. He was the youngest of his class, but graduated a very close second. He did not believe in spending four years at college, so went through in two. He was a favourite of all who knew him and is missed especially by those who are in the habit of attending the Gymnasium. He is at present attending Columbia College, N.Y. and will probably ere long become famous as a scientist.

How prophetic were those final words, written a century ago!

CHAPTER 4

The School of Mines

*T*here were several reasons for Will Matthew's decision to attend the Columbia School of Mines. First, he definitely wished to live the life of a geologist, which is not surprising, when one considers that he was the son of a man who had gained considerable fame in the geologic profession. Second, he had an unusual background of actual geologic experience—the result of tutelage from and assistance to his father. And finally, as he frankly admitted in later years, he had visions of becoming a mining geologist, thereby "making the family fortune." He had seen all too much of his family struggling to make ends meet. For him the Columbia School of Mines was a logical choice.

The school, established in 1864, was the first such school in the United States and was preeminent among mining schools. It was founded by Thomas Egleston, who had studied at the École des Mines in France and thus brought to Columbia the French tradition of scientific education. The school was a beacon beckoning to Will Matthew.

Perhaps it was ironic that he, a descendant of Loyalists who had fled New York in 1783, should return to New York a century later to attend what had once been a royal college. Before the American Revolution Columbia had been King's College, operating under a charter granted by George II, king of England. (It was one of three royal colleges in the colonies, the other two being Harvard and William and Mary.)

King's College had been founded in 1754 and, for a little more than a century, was located on Park Place in lower Manhattan. In 1857, having outgrown its original quarters, it moved uptown to Madison Avenue at Forty-ninth Street. (It was here that Matthew spent his six Columbia years.)

In 1814 Columbia College had been granted by the State of New York a tract of land that had once been the site of "Elgin's Botanical Garden." Owing to the bankruptcy of Dr. Hosack, the proprietor, it had been sold, in 1811, to the state. This property, perspicaciously retained by Columbia until its sale in 1985 for $400 million, is the site of Rockefeller Center.

Although some plans called for moving the college to this site, the move was never made. Instead Columbia College acquired in 1894 the Bloomingdale Asylum grounds at Broadway and 116th Street, the present location of Columbia University.

The cornerstone of Low Memorial Library was dedicated on December 7, 1895, and in 1896 the trustees authorized Columbia College to become a part of "Columbia University in the City of New York." The status of the university was established by the state legislature in 1912, and thus Columbia College, in 1865 a small mens' college of 150 students, grew to become the university that today dominates Morningside Heights.

When Matthew arrived in New York in 1889 Columbia was an institution in transition. It was officially a college, but it was more than a college, for there were 1700 students and a faculty of 203, distributed among the Arts College, the School of Mines, the School of Law, the School of Political Science, and the School of Medicine— officially, the College of Physicians and Surgeons. Also there was Barnard College, the womens' undergraduate college recently founded as a counterpart to Columbia College, which then admitted only men. (Barnard at its beginning accomplished something long desired by Columbia College, but as yet not achieved—a separate budget and a separate board of trustees, albeit both Columbia and Barnard colleges had their own faculties.) Also on the site is Teachers' College—in

1889 only recently established—which, although also a part of Columbia University, has its own budget, its own faculty, and its own president.

One can see that when Will Matthew came to Columbia it was already a complex institution, on its way to becoming an even more complex institution.

Will made the journey from Saint John to New York by boat, as was usual in those days, and after landing in the great seaport crossed over to Brooklyn to "Aunt Nellie's house," where he was to live for several months. Aunt Nellie was a sister of Will's mother; born Ellen Caroline Diller, she had married Harry F. Wilson, a New York stockbroker.

Soon after he had settled in the Wilson home he again crossed the East River and from lower Manhattan went up to Columbia College to apply for admission. There were some worrisome hours before he finally was admitted to the School of Mines.

> I went up to Prof. Van Amringe [Dean J. H. Van Amringe, one of the great names of Columbia history, a man of whom it has been recorded that "few men associated with the college, perhaps none, had such a gift for arousing lifelong affection."][1] and presented my diploma and he told me that I was admitted to the first class. . . . So I went away satisfied. But next Monday morning, I found I was only admitted in three subjects. So I went to the examination in English and explained my case and asked him if he would admit me without a conditional, for free students are not allowed conditional exams. He consented, and then I went around to all the other examiners, asked them to do likewise, and all of them admitted me. Last I went to the Drawing Examiner, Prof. Trowbridge, and explained to him the way I was situated, he was kind enough to admit me on my giving him a specimen then of what I could do. So I was all right, and I presented my admissions to the registrar and finally got my ticket on Tuesday.
>
> (letter to his mother, October 13, 1889)

Having become officially entered in the institution, Matthew began a course of study that was sufficiently rigorous to occupy almost all his waking hours—physics, chemistry, mathematics, drawing, blow-

[1] Horace Coon, *Columbia, Colossus on the Hudson* (New York, Dutton, 1947), 85.

pipe analysis, and botany being prominent in the schedule of his first year. Some of his instruction was received from notable professors: Van Amringe—mathematics; Chandler—chemistry; Egleston—mining; Moses—mineralogy, to mention a few. These subjects, taken during his first year, give some indication of the breadth of Matthew's education at Columbia. In later years he had courses in optical petrography, assaying, surveying, zoology, and paleontology, as well as other courses in the earth and life sciences. If this seems a strange mixture of topics for the School of Mines curriculum, it should be explained that during the latter part of the nineteenth century the school was the fountainhead of all science study at Columbia.

"Almost without exception the present university departments devoted to the natural sciences originated in the old School of Mines."[2]

So Will Matthew labored mightily during the autumn and early winter of 1889, commuting every day between his Aunt Nellie's house in Brooklyn and Columbia in Manhattan. Sometimes he crossed the river on the Brooklyn Bridge and sometimes he used the ferry. The former route was quicker; the latter, cheaper—it saved him two cents per trip, each way. If such a saving seems inconsequential, it should be remembered that the cost of many items are now many times greater than they were in the final decade of the nineteenth century—perhaps by a factor of twenty or more. Thus, two cents per trip is equivalent to forty or fifty cents or more in today's terms—a saving to be seriously considered by a very impecunious student.

Christmas came and Will went home, by boat, for the holidays. On the return trip to New York he disembarked first in Boston and visited Harvard College where he

> took a walk around, and saw the centre of the universe and the common thereto adjoining, and a number of monuments, one of which showed the British Lion being squelched by the American Eagle.
>
> When I reached Brooklyn I found that Aunt Nellie had already appropriated my room and broken open my trunk and packed most of my things in it. So I had to go over to Mr. Pike's [a good friend] and he went around with me till we struck this place. [137

[2] Coon, *Columbia,* 254.

East Sixteenth Street, between Irving Place and Third Avenue, now the site of Washington Irving High School.] I have a little hall room, and pay $6 a week. A Piano, a banjo and 450 cats are extras—no charge for them though.

(letter to his mother, January 8, 1890)

It is hard, even a century later, not to feel a sense of dismay at Aunt Nellie's high-handed behavior. Without discussion or any warning she dispossessed her quiet and studious nephew from his lodgings; consequently, on the spur of the moment he had to find a place to live— on a cold January day. Yet he did not seem to harbor ill feelings against Aunt Nellie; his letters indicate that he occasionally visited and dined at the Wilson home in subsequent years.

Matthew's sudden move to East Sixteenth Street was the first of a series of changes in lodging that continued through his six years at Columbia. Later in the spring of 1890 he moved to 148 West Forty-sixth Street, between Sixth and Seventh avenues—now the location of the Celanese Building—to share a room with a college friend, Will Pomeroy. From then on he made yearly moves, finding lodging in the west Forties or on Lexington or Amsterdam avenue. Most of his accommodations were dismal enough, but in 1892 he had the rare experience of living in a boardinghouse filled with congenial people. It was at 238 West Forty-third Street, in the Times Square area (across the street from the present location of the New York Times plant and the stagedoors of the Lyric and Apollo theaters). Here the lodgers took pleasant meals together, and held Sunday afternoon teas and musical get-togethers.

His subsequent years at the School of Mines were similar to his first, with variations. He gained confidence as time went on and, although his studies were always demanding, there were more opportunities for recreation and relaxation. He made frequent use of the college gymnasium, he went to occasional social affairs, and now and then he attended the theater.

Always, however, he was plagued by his stringent financial situation. He never had enough money, even though he practiced economies that bordered upon penury. Certainly his Scotch-Canadian background gave him ample experience in how to live on very limited resources.

It costs me about 10 cents a day for lunches—I have to get something which is substantial and yet can be eaten in 5 minutes, and is *clean*. And you see that there is not much margin left on the $30 a month. For counting 4-⅓ weeks to a month, there is only about a dollar and a half margin left. This month, owing to extra expenses, I am going to be behind hand and will have to ask Father for $3.00 to keep me going. I enclose a copy of my exp. for the month that you may see what the extras are.

(letter to his mother, January 19, 1890)

Will's handwritten account of expenses is reproduced here (See figure 8). This lists the "extras" that concerned him, such as the boat ticket, laboratory supplies, and eighty-five cents, paid on January 8, for having his trunk moved (thanks to Aunt Nellie) from Brooklyn to East Sixteenth Street in Manhattan. He seemingly borrowed thirty dollars from Mr. Pike—the person who helped him find his first Manhattan lodgings after Aunt Nellie's banishment.

The ledger is of particular interest because of the light it throws on the value of money a hundred years ago. Almost all the items listed are less than a dollar, and even the largest expense, six dollars for a week's board, translates to less than a dollar a day for food. In today's inflationary environment such prices are scarcely believable, but for the evidence set down in Will's neat hand.

Some two years later the money problem and the routine of his studies were getting under his skin.

I wish I were through here—I am getting very tired of doing nothing but spend money. This is much the most unsatisfactory year I have had; the subjects are too large and too many for me, and really I don't think they are very necessary for me. What earthly use, for instance is it for me to learn about 'Electrical Measurements.' And the Mining and Metallurgy are the same way, though to a less extent. If I could spend my time at learning something of Zoology and Botany I think it would be much better; but these are made altogether subordinate subjects. The Zoology that I got last year was by no means in my course—I was supposed to have Mechanical Drawing—only I kicked. But the Metallurgy is run by Dr. Egleston and it is no use kicking about it—he is as crossgrained and cranky as ever man could be.

(letter to his mother, December 13, 1891)

Perhaps there is more in this complaining letter than is immediately apparent. Obviously, he was less than happy as an Egleston student (in a later letter he still dwells upon his dissatisfaction with Egleston and with metallurgy). Beyond that, are there not clues that he was developing an increasing interest in animals and plants—past and present? He was already studying botany, and it is probable that he had become acquainted with Bashford Dean, under whose guidance during the next academic year, he was to study vertebrate biology.

In October 1891, about two months before he wrote that discontented letter, he first met John Strong Newberry, who was making one of his last visits to Columbia. Matthew reported to his father that "Dr. Newberry is around college again. I was formally introduced to him the other day—he spoke of having heard of me as your son, etc. He is not at all strong yet—and I am afraid, never will be again—and is quite unable to teach" (letter, October 25, 1891).

Newberry, a bearded, patriarchal figure, under whom Bashford Dean had studied, is famous in the annals of North American geology. He was an authority on fossil plants and fossil fishes, and he had participated in some of the early natural history surveys of the West, before the Civil War. (Having been trained as a medical doctor, he was later in charge of the "Sanitary Commission" for the Union forces during the Civil War, in the western theater of operations.)

Newberry died in December 1892, a little more than a year after young Matthew had been introduced to him. But is it possible that during the interval between October of 1891 and December of 1892 Matthew became inspired by Newberry's paleontological accomplishments?

Perhaps, since during the autumn and winter of 1892 he was, as has been mentioned, a student of Bashford Dean. He must have had some of Newberry's special knowledge of fossil fishes passed on to him by Dean, to supplement the overall view of fish evolution he was acquiring under Dean. Further, Dean himself must have had considerable influence on the paleontological outlook of Will Matthew.

Bashford Dean was, during the latter years of the nineteenth century and the first two decades of the twentieth, a world authority on the evolution of fishes. His particular interest for many years was in the armored fishes of early and middle Paleozoic age. Some of these fishes, such as the Devonian armored fish, *Dinichthys* (in technical terms, an arthrodire), were gigantic—twenty feet or more in length. Dean became interested in the mechanics of articulating and overlapping armor plates and decided to study medieval armor.

(As a result of this interest in armor, years later Dean crossed Central Park in New York, from the American Museum of Natural History, where he was curator of fossil fishes, to the Metropolitan Museum of Art, where he became curator of armor. There, with the munificent support of J. P. Morgan, he assembled one of the world's greatest collections of medieval armor. It was a remarkable transformation of scholarly interests and activities.)

In 1891 Seth Low was elected president of Columbia by the trustees, and immediately the old college entered upon a new life. It soon moved uptown to Morningside Heights to become Columbia University, but during the interval between Low's taking office and the move changes were made. For example, in 1891 Low abolished compulsory chapel (Columbia was originally a Church of England college). More pertinent to Will Matthew, in 1891 Low brought in Henry Fairfield Osborn, from Princeton, to found a separate Department of Biology. At the same time Osborn founded a Department of Mammalian Paleontology—soon to become the Department of Vertebrate Paleontology—at the American Museum of Natural History.

These developments were happening under the nose of Will Matthew, and of course he was very aware of what was going on. In 1893 he wrote his father

> They have a pretty good library over at the Bio. Lab., the gift of Chas. H. Senff, as a memorial to Dr. Northrop. The department as a separate one, is new, and is well equipped both as to men and materials. Dr. Osborn is at the head of it, and there are also an assistant professor, an instructor and a tutor. Dr. Wilson, Dr. Dean and Mr. Willey.
>
> Columbia is almost the only great Eastern college that has not lost professors owing to the beguilement of Leland Stanford University and the University of Chicago. I suppose it is because they pay their professors well—a full professor here gets from $5000 to $10000 a year, I believe—and also because there is every prospect of a rapid development of the college in the next ten years.
>
> (letter of February 26, 1893)

Osborn and the American Museum of Natural History were to be dominant factors in the life of W. D. Matthew, as shall be seen.

An important requisite in the education of a geologist and a paleontologist is fieldwork. Therefore, advanced students traditionally spend their summers in the field, more often than not as assistants to their professors or to other professional earth scientists. Thus they study geologic structures and the succession of rock strata, and collect rocks and minerals and fossils under the trained and critical eye of someone who has done it before. The necessity of such field training cannot be overemphasized. Without the firsthand experience of tramping across the landscape, of trying to decipher the geologic puzzle of the surrounding scene, of breaking rocks with a hammer and digging out fossils, the student of earth science will have no sense of its reality.

Will Matthew had enjoyed an unusual introduction to geology in the field when, as lad, he accompanied his father on excursions through the New Brunswick countryside. Now, as a young man with classroom knowledge recently acquired, he was naturally eager to get some solid field experience. In the spring of 1891 he had applied for and obtained work as a summer assistant on the Geological Survey of New Brunswick—this after applications to the New Jersey and New York surveys had been unsuccessful. He even arranged to have his examination in Zoology set forward so that he could get an early steamer to Saint John. It appears that he put the summer to good use.

Then during the academic year of 1891–92, he became a dedicated disciple of Professor James Furman Kemp, Newberry's successor, and the beloved "Uncle Jimmy" to generations of geology students at Columbia. In time Matthew became an assistant to Kemp, helping him lead students on field excursions and working for him in the laboratory.

It is therefore not surprising that he spent both the summers of 1892 and 1893 doing geological field work in the Adirondacks, in part with Professor Kemp. This was in New York on the west side of Lake Champlain, where the geology is in many places exceedingly complex, and where important fossils can be found.

> At Port Henry Dr. Kemp arrived on Saturday afternoon, and since then we have been travelling around in a carriage, mostly getting dips, strikes and levels in the rocks of Moriah Township,

nearly all Archaen, mostly hornblende gneiss, or garnetiferous gneiss, with some limestone, and beds of iron ore, several of which are largely worked.

The country is hilly and rugged; it is in the foothills of the Adirondacks, and whenever you get on a hill or rise you see them looming up to the westward. Eastwardly you look across the lake to a broad, level stretch of Palaeozoic limestone, and behind that the Green Mountains stand, quite dim in the distance, stretching away north and south till they fade out of sight. The views this way are really superb.

We went over today to Fort Frederick, at a narrow part of the lake, about two miles from here. . . . There is Trenton Limestone there, full of fossils; we broke out a few. . . . By spending a day or two there one could get a very tolerable collection, I think.

(letter to father, August 28, 1892)

And again, in 1893:

I came up here [Crown Point, New York] Thursday before last, and for a week worked by myself, and had to work on foot mostly, as the working with a horse is not very convenient for only one person. It came to be pretty hard work towards the end, as I had to tramp twenty miles or more a day—not counting any excursions to a distance from the roads—and that is pretty tiresome when one has a bag of fifteen or twenty pounds of rocks to carry. Prof. Kemp and Pomeroy [Will Pomeroy, his classmate] came on Friday.

. . . I am marked "O.K. for graduation." . . . My diploma ought to arrive some time this week.   (letter to his father, June 11, 1893)

This would have been his bachelor's diploma from Columbia.

The Adirondack fieldwork lasted well into September of 1893, so that Will Matthew arrived at Columbia after the beginning of the semester. Upon his return he found a room where he decided to set up housekeeping—a departure from the boardinghouse on West Forty-third. This was viewed with some trepidation by his Aunt Mary over in Brooklyn—she foresaw a sketchy life-style for her nephew—but he went ahead.

The room I have now is a very small one—a hall bedroom, about 6 ½ by 10 feet. It has a closet with running water and another for clothes, also a fire-place where a grate was intended to be, and an oil-stove is. I have got quite a lot of furniture in it, though, a folding bed, wash-stand, folding chair, straight chair, $5' \times 3'$ bookcase, $6' \times 3\frac{1}{2}'$ dish closet and $2' \times 2'$ table. The fireplace is screened by a little moveable screen so that the room doesn't look so badly, though pretty well crowded. I got my lamp, and almost all of the dishes I want, and hope to make out very well.

(letter to mother, October 3, 1893)

Matthew was now a full-fledged graduate student, and a member of the recently established Graduate Club.

Last Wednesday the Graduate Club were invited to a meeting at President Low's house. We mustered there about twenty-five strong, and heard a very good talk from 'Prexie,' and afterwards partook of 'feed' and cigars, and had an exceedingly pleasant evening. . . . The nature of the Club I may have told you of; it is composed of all graduates studying for M.A. or Ph.D., and is entirely social in its object. (letter to mother, November 28, 1893)

It has been mentioned that Seth Low brought Henry Fairfield Osborn to Columbia from Princeton in 1891. By the time Matthew began his graduate studies, Osborn had established the Biology department at Columbia, and was teaching courses in vertebrate paleontology. In the fall of 1893, Matthew was succumbing to the Osbornian influence, but it is difficult to know just how much Matthew was affected by Osborn's teaching, or how seriously he took the subject of fossil vertebrates. He still looked to advance in the field of hard rock geology, yet his interest in the bony remains of ancient animals loomed in the background. His plans and feelings are indicated in letters he wrote to his father in the autumn of 1893: "My major, as was arranged some time ago, is on the granites, etc., around St. John; my minors are, in Mineralogy, a general systematic course with special reference to one group, the amphiboles; in Biology, two courses, one a general preparatory practical one, the other a course on Mammalian morphology" (letter of October 15, 1893). And, "Why did I take the course in Vertebrate Osteology? Simply because it was the only thing in the line of paleontology I could get at the college. The only other courses, in Palaeobotany and Fish, I had had" (letter of November 28, 1893). And then this rather surprising statement: "I am working along on the Biology course, at the osteology of the Mammalia. Prof. Osborn seems to think I am getting along pretty well, and as he is the most influential man on the Pure Science Faculty, I want to keep in with him as much as possible with a view to re-election as Fellow next year. *I don't get up an intense enthusiasm over bones, however, tho' it is a fairly interesting subject* (letter of November 14, 1893, emphasis added).

There is no inkling here that the young man who wrote these words was destined to be one of the greatest students of fossil mammals in the history of paleontology.

Speaking of destiny, Matthew and Osborn were to be closely, if not always harmoniously, associated for more than thirty years. Thus, in any account of Matthew's life and career there is much to say about Osborn, a complex, protean man.

Henry Fairfield Osborn was born in 1857, fourteen years earlier than Matthew. He was a child of wealth. His father, an important figure in the world of business—having been the president of the Illinois Central Railroad—was one the several nineteenth-century businessmen who amassed considerable fortunes and then lived in late-Victorian splendor. He built a veritable castle high on a hill overlooking the Hudson River, north of New York City. This dwelling, known as "Castle Rock," faced West Point across the river and was a prominent landmark of the region.

Henry Fairfield Osborn (he was always proud of the full name) went to Princeton University, where he became a close and lifelong friend of William Berryman Scott. Osborn (known to Scott as "Polly") and Scott (known to Osborn as "Wick") were intended by their families to go into the worlds of business and railroads, in Osborn's case, and medicine, in Scott's case. But one spring day in 1876, just before the examinations at Princeton, they decided that it would be fun to go out to the wild and woolly West to hunt fossils. (This was the year of the Battle of the Little Big Horn.) Indeed, the next summer they did go west, to the Bridger Basin, where, with a friend, Francis Spier, they collected fossils of Eocene mammals. That settled the fate of Scott and Osborn—they both became paleontologists of renown.

After graduating from Princeton each went to Europe for postgraduate studies. Osborn was a student of Thomas Henry Huxley and once had the never-to-be-forgotten experience of meeting Charles Darwin in Huxley's laboratory.

Both came back to Princeton, where they had been appointed to the faculty. Scott spent the rest of his life at Princeton, but Osborn, as recounted, was lured to Columbia in 1891 by Seth Low.

Osborn was by then in his mid-thirties, well-established as an accomplished paleontologist and known as a forceful man with ideas—some good, some debatable. At Columbia and, simultaneously, at the American Museum of Natural History he began to develop ambitious programs of paleontological research, his efforts at the museum being directed toward building up collections of fossil vertebrates.

It was at this point that Will Matthew, barely beyond his majority, entered the Osbornian world.

Just as Matthew's career was to be unalterably affected by Osborn's arrival at Columbia, so was Osborn's scientific life to be affected by the appearance of this quiet young man in his classes, commencing in 1893. Of course, in the fall of 1893 neither could foresee how their lives were shortly to become intertwined. Osborn—tall, imposing, and quite confident of his proven abilities—seemed the very antithesis of Matthew—slight, spare, and still groping his way toward an uncertain future. Yet the scientific lives of professor and student were to be written large together in the annals of American paleontology.

In those first years of their acquaintance Will Matthew was proceeding along what he considered as his predestined course.

It will be remembered that at the beginning of the fall semester of 1893 Will had settled into a remarkably small room, where he intended to keep house and do his own cooking. Evidently that was a short-lived experiment; he soon moved into a much larger room shared with a fellow student named Gregory (unrelated to his future paleontological colleague, William King Gregory). Just before Christmas this room received the attention of a thief.

> We had to change our quarters, as things were beginning to disappear too fast at 336. Gregory lost his best coat, and I lost a present which Aunt Chrissie made for me. So we pulled up stakes as fast as we could. We could not get any clue as to the thief, and the police and detectives would do nothing and told us that we could do nothing. We are now boarding, and close by the College,

so that we are not cut so short as to time. . . . The place where we are now [626 Lexington Avenue between 53rd and 54th streets, across the avenue from the present site of Citicorp Center] is a very neat and pleasant room and promises to be very satisfactory.

(letter to his father, January 31, 1894)

Despite the unpleasant experience, Matthew was not particularly disturbed. He went to the theater with his cousin "Alfie" Diller with whom he had played in Brooklyn when they were both little boys, and he spent the Christmas holidays in Philadelphia with Uncle Rob, Aunt Chrissie, and a house full of cousins.

He made plans during the winter for the coming summer field campaign. Professor Kemp had promised a small contribution from departmental funds to help Will on his contemplated project, which included geologic mapping in new Brunswick, as well as collecting rocks and fossils. As late as March he was planning to work on the Cambrian rocks there, with a classmate named van Ingen, who was to become an outstanding paleontologist.

Then in April his fellowship was renewed—a pleasant surprise, because in March he had considered this an unlikely event since "the representation given to Pure Science will probably be considerably reduced, spite of Prof. Osborn's influence" (letter to father, March 17, 1894).

But Professor Osborn had made his influence felt in a very positive manner so far as Matthew was concerned. Indeed, it seems obvious in hindsight that Osborn had been regarding with increasing interest and favor the student from New Brunswick who, for more than four years, had supposed his future to be in mining geology.

Will Matthew's life took a turn during the late spring of 1894 that can be largely attributed to the insight and the wishes of Professor Osborn. Not only was the fellowship renewed (Osborn's insight) but also, as soon as the semester was over, he journeyed to North Carolina and went into a subsurface coal mine (Osborn's wish). Ironically, Matthew's descent into the mine had nothing to do with mining geology, for which he had assiduously prepared himself through four years of intensive study, but rather with the collecting of fossil vertebrate animals, to which he was a recent initiate.

To understand the sudden change of direction that Matthew was taking—from a full summer along the North Atlantic seaboard to a partial summer in a humid Carolina mine—one returns to 1856. It was then that Ebenezer Emmons, a pioneer North American paleontologist, described two tiny fossil jaws (each less than an inch in length) found in the black Triassic coals of North Carolina. He judged these jaws, which he named *Dromatherium sylvestre,* to be of particular importance, because they seemed to him to be the jaws of the earliest-known mammals. Such was also the opinion of other knowledgeable paleontologists, including Osborn. Osborn was intensely interested in the origins of mammals, so he studied the two specimens, one at Williams College, the other at the Academy of Natural Sciences in Philadelphia. Osborn decided the Philadelphia specimen was different from the Williams fossil, so he renamed the former, designating it as *Microconodon.*

In 1894 no additional specimens of *Dromatherium* and *Microconodon* had been found, so he sent Matthew to North Carolina to search for fossil protomammals. (It is now recognized that the two tiny jaws are not those of mammals, but rather of mammal-like reptiles, from which the mammals arose.) Matthew labored hard in the coal mine but found neither *Dromatherium* nor anything like it. Indeed, to this day no more *Dromatherium* or *Microconodon* have come to light.

He did, however, discover in the black shales and coals of the mine numerous bones of a phytosaur—a late Triassic crocodile-like reptile. A skull of such a phytosaur had been previously described by Emmons, who named it *Rutiodon carolinensis.* So Matthew spent his summer collecting phytosaur bones.

> I get in a pretty long day of it, as I am generally not in to supper till after sun-down, for it is a long business to clean up after a day in the mine. I had to get a suit of overalls, for the old suits I took with me are not old enough, and do not keep out the wet, either. And over those I wear an old rubber overcoat while going down the shaft. This is a very wet mine in the shaft; the workings are fairly dry.
>
> No *Dromatherium* has shown itself yet, but I have found a large number of bones of Saurians and any quantity of their teeth and of scales of ganoid fish. There is scarcely anything to be found in the way of vegetable remains, only a few structureless bits of lignite in the coal; and no insect remains. The saurian bones appear to be ribs, leg and skull-bones, and bony armour plates, and others which

I do not recognize. The teeth are of several types, but none mammalian. Altogether I have not been very lucky, though I have broken up about 1000 square feet of the coal bed in which the remains lie. I may find the little rats yet, though, before the week is over. . . .

Well, I am getting a good deal of information of various kinds here, even if I am not finding any fossil mammals. And I have partaken of 'turkle' stew, the meat of which animal tastes in different parts like chicken, beef, pork and veal respectively, while the bones are very like the saurian bones in the coal. Which fact one of my miners commented on today. They are both intelligent fellows and good workers.

(letter to his father, from Egypt, North Carolina, June 19, 1894)

After his work in the coal mine he did go to New Brunswick, where he collected Cambrian fossils, and in September Will was back at Columbia to complete his sixth year with the aid of his fellowship. He was busy with his final graduate courses and his thesis, and—although he was continuing his vertebrate studies under Osborn—his letters are devoted in considerable detail to the Paleozoic rocks and fossils of New Brunswick. His predetermined thesis problem under Kemp's direction was maintained despite his new involvement with fossil bones.

On April 16, 1895, he wrote to his father: "My thesis is in the printer's hands and I expect to have it out soon." The thesis, published in *Transactions of the New York Academy of Sciences,* was entitled "The Effusive and Dyke Rocks Near St. John, New Brunswick." And with this publication, his eighth, he effectively bid farewell to the study of the Paleozoic volcanic and fossiliferous rocks of New Brunswick. Except for two later short contributions on New Brunswick geology, one on volcanic rocks (published in 1895) and one on intrusive rocks (published in 1897), he never again published anything pertaining to "hard-rock geology." Henceforth his scientific life was devoted to fossil vertebrates, principally mammals.

His letter to his father of April 26, 1895, announced "you will already have heard of my appointment to a position at the Amer. Mus. Nat. Hist. . . . My salary will be $80.00 a month, not far from a thousand a year." So the die was cast that would lead to a future in paleontology involving more than thirty years at the great museum in New York.

But in the immediate future was another summer in the North Carolina coal mine.

> We are in for the hot weather now, and with it at least one of the seven traditional plagues—namely of flies. I never saw such swarms before, even last year while I was here. It is fortunate that I can get away from them for the best part of the day, at least, by going down the mine.
>
> I have been making great hauls in the line of reptiles. One skull which I have got out is 31 inches long, with its great long jaws gaping apart like the old Ichthyosauri you see in all the books on geology. There is a great quantity of other bones, all separate except a few which still cling together in their proper places. This is far ahead of my finds last year in every way. . . . If only I could find some mammals now, I would be completely successful; but the little rascals are not to be seen anywhere, and I'm afraid it is a hopeless task to look for them. . . .
>
> This mine work is about as unpleasant as any I know of. It is so abominably dirty; and there is no place to bathe except the muddy and clay-bottomed Deep River. That's better than nothing, though. It is very cramping work, too, in the narrow and low-roofed galleries. I have to lift the slate floor and work in the coal underneath. Of course it is a great alleviation to find these fine bones, but I won't be sorry when it is over and done with. . . .
>
> My name has been approved for doctor's degree, so I may expect to have the right to add another tag to my name in a few days. . . . I will send a copy of my thesis to Father as soon as I get any.
>
> (letter to his mother, from Egypt, North Carolina, June 15, 1895)

Some of the bones that Will Matthew collected in the North Carolina coal mine have been assembled as a skeletal mount of *Rutiodon carolinensis* at the American Museum of Natural History (see figure 11).

CHAPTER 5

The Young Paleontologist

*B*y early autumn 1895, when Indian summer engulfed New York City and when, across the broad Hudson, the first tinges of color had touched the woodlands that crowned the New Jersey Palisades on the opposite shore, Will Matthew had settled in at 509 Amsterdam Avenue, between Eighty-fourth and Eight-fifth Streets, with his brother George. It must have been a memorable time for the young paleontologist, principally because he had crossed two thresholds in his developing career. He had obtained his doctorate, so his life in academia was behind him; he had been appointed to a staff position at the American Museum of Natural History, so his life as a professional scientist was now before him. Those fall days must have been psychologically exhilarating as he looked toward his future in paleontology, and they must have been physically exhilarating as he felt the cool, crisp air that can make that time of year in Manhattan quite delightful. It was a contrast, indeed, to the heat and humidity of the Carolina mine.

That September Will Matthew had something to think about in addition to paleontology—the self-imposed obligation of getting his brother George initiated to and established in an academic life at Columbia. It was a part of the Matthew family tradition that, as the eldest son, he assumed responsibility for his younger siblings, especially for the brother next in line.

On April 26, 1895, Will had written to his father, outlining a plan for George to come to New York.

> You will already have heard of my appointment to a position at the Amer. Mus. Nat. Hist. I write to you to consult you about a plan which I have been thinking lately, and I want to get your opinion of it. My salary will be $80.00 a month, not far from a thousand a year, and I have been figuring on the practicability of George coming on and going to college at Columbia next fall. I think I could just about cover our joint expenses for the first year, and after that I would hope that he could get something to help along in the way of tutoring or other work. . . . It will be some extra work if George comes, as it will involve my keeping house, and some penny-saving that I would not otherwise find necessary. But I can balance his being here against that, and come out ahead.
>
> In case you think this too close shaving, it might be possible for George to come on here and get some work. . . . Of course it would be a great pity for him to give up all idea of going to college; I mention it only in case of there being no way out otherwise.

Evidently the contemplated shave was not too close, because George came as a prospective student—to share his older brother's apartment and eighty-dollar monthly income. However, some problems arose as a result of the difference in the personalities of the brothers. Will was studious, devoted to his science, and methodical. George was something less than a dedicated scholar, devoted to music (as were all the Matthew siblings, except Will), and a bit haphazard when it came to keeping accounts. Within a few weeks, the elder brother was expressing some doubts about his housemate. (We do not have a record of his thoughts, but perhaps the younger brother had his own opinions.)

> I have grievous fears that George will not succeed in making up his deficiency, or not with credit enough to get free tuition next year. I don't know of course what amount of concealed genius he may have, but it's commonly reported that the majority of men who do well at college do so by dint of continued study and hard work.

I should be not a little mortified if he should fail to make up his conditions, for I feel more or less responsible to Prof. Kemp in the matter, and Prof. Kemp has stood surety for George in his getting in. I had a line from the Professor, in which he implies (in his usual kindly and considerate way) that he expects George to "locate himself in the room at the top" as a return for the Faculty having granted him the favor. So that I hope he will see fit to drop most if not all of his music, and settle down to work.... And if George will only do himself credit at College and not fritter away his time on music, and visits to Brooklyn, he will be able to get a scholarship next year, and we can then get a little ahead.

(letter to mother, November 4, 1895)

There are more complaints in the letter—about George failing Greek and Latin and about his sloppy household accounts. Perhaps Will was too harsh with his brother; how can one censure a love of music? Music was a very integral part of George's liberal education. And, as it turned out, George's liberal arts education paid off very well indeed; in later years he founded The Tutoring School of New York in Manhattan, an excellent and respected tutoring school, still in existence, over which he presided for many years.

But of George's subsequent success there was as yet no clue, and so by the end of the academic year Will was still worried about George's performance and prospects. Evidently George had not particularly distinguished himself as a first-year student and was thinking about trying something different. Will wrote his father on June 18, 1896, with some suggestions:

I should think that George would make a very serious mistake in taking a special course at Columbia, and waiving his chances for a degree. He would do far better, if he cannot make up his mind to get through his quota of classics, to take a full course at some other college.... What would you think of George's dropping his college course entirely, and studying music as a profession? [So! the elder brother belatedly recognizes that music has its merits] I could do as much for him as I have been doing—videlicet, keep on the present arrangement as to living, and pay something towards his tuition expenses, whatever they might be. He will have to drop either one thing or the other, if he is going to do himself any credit here.

Perhaps Will's concern with George's problems was exacerbated by the pressures of his work at the museum. A new member of the staff, he was keenly aware of being in an environment where he had various responsibilities. Almost overwhelmed by fossil collection duties, he was also engaged in research under the watchful eye of Jacob Wortman—and in the background were the insistent demands of Professor Osborn.

During the winter of 1895–96 Will was struggling with a full agenda at the museum, consisting of four principal projects, each of which might have taken all his time and energy had he not rationed his hours and programmed his activities with a stern hand. These projects were cataloguing the Cope collection; revising the Puerco fauna—this being partly related to his work on the Cope Collection; classifying the freshwater Tertiary sediments of western North America; and organizing materials for an exhibition hall.

With the last of these duties, he was under considerable pressure because of a looming deadline. (This is a recurring complaint of museum people: exhibit dates seem always to intrude into a curator's program, superceding all other plans.) On January 30, 1896, he wrote his father: "The Museum work goes on much as usual, but I am pressed a great deal for time at present, as the new hall is to be opened about Mar. 1, and I ought to have the greater part of Cope's Collections catalogued by that time, which I have some doubt of doing unless they are more prompt in filling out orders for supplies than experience warrants me in expecting." There was probably frantic activity in the last days of February.

Regarding the Cope Collection, some discussion is in order.

Edward Drinker Cope, scion of a prominent Philadelphia family, was one of the pioneer naturalists of nineteenth-century America. He was a man of exceeding brilliance—one might say he had more than a touch of genius—with a considerable inheritance at his disposal. He devoted his life to collecting and studying fishes, amphibians, reptiles, and fossil vertebrates. His collections were extensive (he hired collectors to search the western badlands for fossils), and the sheer volume of his published studies of the materials he acquired was prodigious: the bibliography of his works runs to more than fourteen hundred titles.

Before Cope's death on April 12, 1897, the officers of the Philadelphia Academy of Science had logically, but mistakenly, assumed that the Cope Collection—at the time stored in two adjacent town houses on Pine Street—would be willed to the academy. They were not. Osborn, a Cope "disciple," had, in 1895, persuaded his uncle J. P. Morgan and the other trustees of the American Museum of Natural History to purchase the Cope Collection of fossils. Some mainline Philadelphians were very unhappy about this fait accompli executed under their noses, but there was nothing to be done.

One of Matthew's early assignments at the museum, after his second summer in the coal mine, was to go to Philadelphia and pack the Cope fossils for shipment to New York. The first Cope fossils to which Matthew directed his attention were those of ancient mammals from the Puerco and Torrejon beds of the San Juan Basin in New Mexico, and this led to his "Revision of the Puerco Fauna," published in 1897. It was his first substantive paleontological contribution, a critical review of sixty-four pages in which he distinguished the successive Puerco and Torrejon mammalian assemblages that lived at the beginning of the Age of Mammals.

This research was on his mind in the early months of 1896—as was his work on the classification of the western Tertiary sediments—and on March 10, 1896, he wrote a long letter to his father about the problems with which he was struggling. The letter ran to six pages and included a chart showing the succession of fossil mammal–bearing sedimentary beds in North America. It is a historic letter, and a particularly historic chart, for it adumbrates the materials that were to appear in Matthew's publication in 1899, "A Provisional Classification of the Fresh-Water Tertiary of the West" (*Bulletin of the American Museum of Natural History,* Vol. 12 (1899):19–79). Today's paleontologists will be interested to see Matthew's chart as he originally drew it for his father (see fig. 18).

At the bottom of the chart Matthew observed that "these characteristic forms are arranged mainly in order of size—partly in order of abundance. You will note that a group climbed to the top of the list (i.e., attains its maximum of size and abundance,) and then disappears. It is the most striking feature of Mammalian succession, and yet I doubt whether it is popularly realized as a fact."

In an earlier letter, written on January 30, 1986, Matthew had told his father about having begun his work on the Tertiary mammal-bearing beds. He predicted that "the classification will be a good one and authoritative, as it is to be approved by Cope and Scott, and will no doubt be the standard one for some time to come." It was.

Significantly, it was the cataloguing of the Cope Collection—first in the glorified rat's nest that filled Cope's two adjacent town houses on Pine Street in Philadelphia, when Matthew was packing the fossils, and then, at more leisure, in the museum laboratory—that established the beginning of Matthew's encyclopedic knowledge of mammals.

He was methodical, he was careful, and he *studied* every fossil as he catalogued it and wrote the accompanying label. And whatever he studied and catalogued he retained in his phenomenal memory.

As winter gave way to spring in Manhattan the young paleontologist was experiencing the undiminished pressure of his work, compounded by a feeling of confinement within a world of brick walls, unrelieved by trees and grass and spring flowers. He needed a release. When he received some money from his parents with which to purchase a bicycle, he wrote a letter of appreciation:

> I don't know how to thank you enough for the wheel. It is the one thing in particular that I had set my heart on having, and yet I couldn't see my way clear to getting it for some time yet—two or three years at the very least, if ever. [Remember, his brother George was with him, and Will was bearing most of the expense of George's education, including bed and board.] I shall enjoy it immensely, for I am especially well placed for using it a great deal, and it will give me the sort of exercise and amusement I most enjoy. I'm looking forward to innumerable trips into the country on holidays, besides a daily ride in Central Park or on the Boulevard or Riverside Drive.... I am going to get a good solid wheel, a "tourist" which I can use for years on country roads.
>
> (letter to his mother dated May 19, 1896)

He got his "wheel," in those days the equivalent of a sports car for the young man-about-town. He did make excursions into the countryside, in the company of friends also equipped with bicycles. Some photographs, evidently taken that summer of 1896, show Will Matthew—neatly clad in men's knickers and sporty knee-length socks (just the thing for riding a bicycle), with a light jacket over a white shirt, a bow tie, and a dark homburg—reclining in the grass by the side of a country road (see fig. 20). That year, after the completion of his graduate studies, he had grown a mustache and beard, neatly trimmed, which added a measure of maturity to his young face. With him are two male companions, and the bicycles in the background are leaning against a tall picket fence, behind which is a wide lawn sloping up to a pillared mansion. On the lawn, its back to the great house, is a Victorian statue of a nymph gazing pensively toward the resting travelers. The scene has all the appearance of Westchester, when that county just north of new York City was still pleasantly rural.

The bicycle gave Matthew the means to escape the city, yet he had an even wider vision of release. In the same letter in which he thanked his parents for making ownership of a bicycle possible, he revealed to his mother his daydreams of a seaside life at the edge of the metropolis.

> Do you know what my last castle-in-Spain is? To have a yacht and a cottage by the Sound, forsooth! And to be managed thuswise. Four of us—say Alf (Alfie Diller), Oscar Blackman (a Brooklyn friend), myself and an Unknown, will buy a little piece of land, about a quarter of an acre, say fifty feet frontage on the Sound and two hundred feet deep, somewhere between Westchester and Greenwich (Conn.). . . . There we'll put up a cottage and get a small cat-boat. And we'll camp out there for the summer months at any rate, coming in to town every day. Now I am pretty sure that in the course of a year after George graduates I can manage to lay aside $500 from my present income; and the running expenses would not be much if at all heavier than mine are at the present, for I should give up the flat, take my furniture out there, and come in to town and board in the winter if I did not like it out there. We would keep bachelor's hall and cook our own meals, of course. . . . We would expect to get a rough piece of land from some farmer or other person out there, and we would grade and level it and fix it up as we had time, making say a little lawn and tennis court in front of the house and putting up a small moveable dock where we would keep a little rowboat, and anchor our cat outside. In the winter we could house it in a small boat-house which would not cost much to put up. Then at the back of the house we'll have a little garden, and a hedge 'round it. . . . It would be immensely jolly to be fixed that way, and go off on cruises on holidays and in the

evenings whenever we felt like it.... The most serious problem really is as to whether there is a piece of land to be had along the Sound—it's nearly all taken up by wealthy people who probably would not sell any part of their grounds.... If the land is obtainable I think I will make a try at it some day, for I am very tired of living in town out of reach of water, green fields or even elbow-room.... If I can't get by the sea-side, I intend to move out somewhere back in the country and buy a lot and build a cottage.

Little did Will Matthew realize, when he indulged in these day-dreams, that a decade later he would indeed acquire land and build a dwelling. But it was not a small plot by the shore, on which there was a cottage inhabited by four bohemian bachelors. Instead, it was a large tract in Westchester overlooking the Hudson River, on which was to be erected quite a sizeable dwelling, to be inhabited by the W. D. Matthew family.

For the present Matthew was to have an uneventful summer in New York until August, when he was to join his family in New Brunswick for a vacation. In writing to his father about his summer plans and the greatly anticipated family reunion in Canada, he dropped an interesting paleontological aside: "Did I tell you that I was up at New Haven a little while ago, and bearded the great Marsh in his den. He was very civil to me, though, and gave me a number of reprints and his monograph on Toothed Birds, and as little as he could manage on the information I was looking for. Which latter was about what I expected" (letter to his father, May 21, 1896).

CHAPTER 6

City and Badlands

*T*he late summer vacation in New Brunswick was for Will Matthew a blessed interval for a renewal of the spirit, a time spent with his family in the place of his childhood. For a few weeks he put behind him the confining brick and mortar of Manhattan, but soon the day arrived for his return to New York, once again to assume his responsibilities at the museum. Early in September he walked into his flat at 509 Amsterdam Avenue; this homecoming entailed a surprise.

When I got here I found the place all fixed up with various strange adornments and conspicuously in their midst an effigy armed with a 'growler' [a 'growler' was, in late nineteenth-century America, a tin-pail filled with beer] and a red flag and with a cabalistic sign the interpretation whereof cast grave doubts on the teetotalism and other high moral virtues of the labelee. Likewise many flippant jests and gruesome puns were scattered about with exceeding recklessness, so that on beholding them I sat down and wept bitterly for the space of about one hour.—I found out afterwards that the perpetrators of this unseemly outrage were a band of wild Dillers (from Brooklyn) who had obtained access to the flat by means of duplicate keys, and were assisted in their horrid work by a black man [his friend, Oscar Blackman, from Brooklyn] and a young person named Raggles [another Brooklynite]. I am still suffering from the effects and have been in consequence unable to write till now. (letter to his mother, Sept. 9, 1896)

After Will had cleaned up the "horrid work" perpetrated by his Brooklyn relatives and friends, he settled in for another year in the apartment, again with his brother George as a roommate. But this year there was a third occupant, his sister Elsie, the eldest of the Matthew children. In the Matthew tradition of family solidarity, Elsie was there to cook, keep house, and "control the purse" (as Will put it) for her two brothers, thereby helping them pursue their chosen fields without domestic distractions. Will certainly had distractions enough at the Museum, and George was still a student at Columbia.

(Elsie was fulfilling a role often assumed by self-sacrificing daughters in Victorian and Edwardian families. She never married. When the two older brothers needed help, Elsie pitched in and provided a pleasant home to which they could return after their days at the museum and the university. Later she went back to New Brunswick to help with the upbringing of the younger brothers. And, after all the other Matthew siblings had left the parental nest, she stayed on to take care of her parents in their old age. Will's younger sister, Bess went to New York after Elsie returned home, and for several years *she* kept house for her brothers. But Bess, who was studying to be a musician, met another music student, Edward Manning, and in 1903 they were married.)

That fall Will was as busy as ever: "Prof. Osborn is back in N.Y. and I am working hard at preparations for opening the Exhibition Hall" (letter of Sept. 9, 1896). Not only was he working on exhibition projects; he was also continuing his research activities, particularly to the Puerco fauna project and his ever-developing classification of Tertiary deposits.

At Christmas he went with Elsie and George to Brooklyn for dinner with the Dillers. The next day the Dillers and Blackmans went on a sleighing party to Coney Island (imagine that!), but Will excused himself because he found the party "a little too gilt edged" for the current state of his wallet.

His holidays were, however, somewhat clouded. He had promised to take part in an amateur dramatic effort along with his cousin Alf

Diller, Oscar Blackman, and other acquaintances. He dreaded the prospect, for he was not an uninhibited, theatrical person. Fortunately, the audience was small—limited to assorted Dillers and their friends.

A more serious problem was an incipient plan for Harrison, the next brother after George, to join them at the Amsterdam Avenue apartment.

> Is there any word or information about Harrison's coming on here? . . . . If H. is coming on I would like to know as soon as possible in order to make arrangements for a change. It will affect my financial plans (I enjoy that word in connection with my large ! salary) [still eighty dollars a month] as I will have to get another bed and probably other furniture, and give notice to my landlord and hunt up a new flat. . . . Ultimately I intend to take a larger flat in any case but will not move for the present unless H. comes on.
>
> (letter to his mother, of December 27, 1896)

Harrison, known to the family as "Hatch," did not come. It was not until a later year that he left New Brunswick for New York, and the same was to the case with his even younger brother, Charles. So Will, George, and Elsie continued to occupy 509 Amsterdam.

In the summer of 1897 Matthew did his first western fieldwork. Osborn had developed a comprehensive program of museum paleontological expeditions, designed to obtain fossil vertebrates for the enhancement of the museum's permanent collections and, at the same time, to inaugurate long-term research on ancient vertebrate faunas and their constituent elements in western North America. Matthew's initial effort in the American West was to be devoted largely to the Cretaceous chalk beds of western Kansas, where fossils of large marine reptiles were to be found. This was not precisely in line with his primary interest—the history of mammalian evolution—but it was high on the list of desideratum in Osborn's scheme of things, and Matthew, as a newcomer to the department, was elected. Perhaps it was a trial run for the young paleontologist—an opportunity to become acquainted with the problems of western paleontological field work.

Therefore, in the spring of 1897 Will was making plans for his venture into the Western fossil fields. It was an exciting prospect, a chance not only to become acquainted with the geology of (and outdoor life in) the region where he was to spend many future summers, but also to prove his mettle in fieldwork—the methods of which are as important to research in geology and paleontology as is laboratory work to research in chemistry and physics.

Yet in contrast the pleasant prospects of the fieldwork were some sobering thoughts about his future at the American Museum of Natural History. Like many a young adult he was feeling some dissatisfaction with his current situation. He appreciated the opportunity to join the museum staff and was grateful to Osborn for having opened a gilded door to a bright paleontological future. Yet he felt the hardships imposed by an eighty-dollar monthly salary. It does seem, even at this distance, a pittance.

Osborn probably was not in any way trying to exploit Matthew by taking him on at the museum at an annual stipend of less than a thousand dollars. Osborn, as noted, was a man for whom money had little meaning. He had never had to live on a salary and, consequently, truly did not know what it was like. Indeed, he considered himself a generous person. He made many contributions out of his own pocket for the advancement of the museum program, especially toward the well-being of his beloved Department of Vertebrate Paleontology. He probably thought that eighty dollars a month was quite sufficient for a young single man just beginning his career.

Matthew thought otherwise, with good reason. Moreover, Matthew saw himself as a prisoner of institutional whims.

In those days there were no such concepts as tenure, schedules for advancement and salary increments, retirement programs, or other institutional practices fairly common today. Unions were things of the future, and arrangements were made on an informal, ad hoc basis, even the planning of a career. One was employed at the pleasure of one's future boss and one remained at his discretion—or at that of the institution itself. So it was with Matthew, Osborn, and the museum.

> In regard to my being open to an engagement at McGill [as had been tentatively suggested by his father] or elsewhere—I must say that in some respects I am not entirely satisfied with my position here, chiefly in that I have reason to fear that unless I can force their hand the Museum authorities are quite likely to keep me on as long as they can on a small salary, not increasing it as time goes on. I know that this method of doing business has been practiced

by them in certain cases which have come to my notice, and have heard that they have a reputation for acting in that way—which I think is hardly good business policy, and certainly not suitable in a scientific institution. At all events, I do not intend to submit to it permanently. For the present, of course, until George graduates, my hands are tied—but after that, if they have not done so of their own accord, I shall ask them to raise my salary and position, and if they do not, will resign and look for something else to do. For the next two years, however, I cannot afford to drop this position unless I got one which would enable me to pay George's expenses to the end of his course at Columbia. . . . I should like very well to be in an institution like McGill, and in my own country. But I can tell much better a year or two years from now.

(letter to his father, April 9, 1897)

Fortunately for the American Museum of Natural History, Matthew stayed for thirty years; he resigned in 1927, not so much for financial reasons as because in part of a deteriorating relationship with Osborn. He truly made his scientific reputation at the museum and in the process did as much as anyone in that institution's history to add luster to its eminence in the field of natural history.

He went on in the same letter to discuss at some length his work on the Puerco fauna.

The Puerco paper progresses slowly—the plan is changed somewhat, and I expect to write the whole of the Revision of the Fauna myself, and have it come out as a separate bulletin. [The original plan was that it be a joint contribution with Jacob Wortman.] I expect there will be a pretty sharp criticism from Prof. Cope when it comes out; he does not at all like the way I have slaughtered his species. But it is a very necessary piece of work if one is to clear up the subject at all—and I think there are a good many formations in which it could be done to advantage. . . .

I expect to run athwart some of Professor Osborn's theories as well. He does not like my arrangement of some of the groups, nor is he entirely pleased, I imagine, that I am adducing some evidence which tells strongly against the theory of foot development which Prof. Cope evolved and he has adopted. So you see I am going to get in hot water. But I have studied my subject pretty thoroughly and believe that my theories, on present information, are most probable; and of my facts I am sure. Dr. Wortman approves of my

grouping, and of the theories as far as I have formulated them—
and I have convinced Prof. Osborn on many points on which he
was at first incredulous.

There was to be no criticism from Cope when Matthew's bulletin
on the Puerco fauna was published later in the year. Matthew had
written his letter on April ninth; three days later Cope died at his
home in Philadelphia.

Less than two months later Matthew was at Cope's Pine Street
houses (it may be recalled that Cope had two adjacent town houses in
Philadelphia; in his last years they served as storage facility, laboratory,
office, and modest living quarters) where he catalogued and packed
portions of the Cope fossil collections. In June, back in New York, he
wrote to Osborn to report what he had been doing.

New York
June 6, 1897

Dear Prof. Osborn:

I have found more snags in the Puerco, and have cleared up
somewhat the character of one more species, *"Chriacus" inversus*
which is shown by additional specimens to have a very peculiar type
of molars, small, short, high cusped, and somewhat insectivore-like.
There is also more material apparently referable to *Microclaenodon
assureus* [another creodont]. I wish I had made a critical study of
the Puerco before cataloguing any of it, instead of relying on Cope's
labels. The result of not doing so is that a large part of the work
has to be done over again, and the numbers are in the worst possible order. I find it necessary to make a catalogue of consecutive
numbers with specific reference of each number, as an index to the
other catalogue. I will not be able to do more than get through
with the Torrejon collections before I leave, and may have to write
up some notes for the paper on the subject while I am out West.

There will be a number of new specimens on exhibition as soon
as they can be bronzed. The hind limb of *Coryphodon anax* [a primitive protodont ungulate] is finished; for the fore limb it would not
be difficult to get a composite specimen, humerus No. 269, radius
and foot No. 258, ulna No. ———, which would fit well together
and make a handsome (?) mate to the hind foot. But as the question
of composites, except as to complete skeletons, has not come up yet,
this is laid over 'till your return.

The Cope notes on N.E. Colorado are missing, unfortunately in
the part relating to the Loup Fork collections; what there is re-

maining I have copied out, and annotated on the margin with iden-
tifications of specimens referred to. The list of papers I have also
got together, and send a duplicate; it is not large but includes all I
have been able to come across.

Owing to his work in Philadelphia on the Cope Collection, Will
Matthew did not get away for his scheduled Kansas field trip until
sometime after the first of June (the date originally set). The area
chosen was in Logan County, in western Kansas, where the Smoky
Hill River cuts a broad valley through the thick chalk beds that char-
acterize this part of the state. These chalks, known in geologic parlance as the Niobrara Formation, were deposited about 125 million
years ago on the floor of a great shallow sea that covered what is now
the high plains region of North America. It was a clear, tropical sea,
unsullied by muds and sands, and in it lived fishes of many kinds,
ancient sea turtles, and various other marine reptiles, among which
were diverse mosasaurs—giant lizards adapted for a completely marine style of life. Today the fossils of these animals are found in the
chalk cliffs, which are frequently eroded into picturesque monuments
and buttes.

Matthew explored the cliffs and gullies along the Smoky Hill River
with Handel T. Martin, a Kansas fossil collector who did much fieldwork and laboratory preparation for the university in Lawrence, but
who, on this occasion, had been hired by Osborn to work with Matthew. By the twenty-sixth of June Matthew wrote to Osborn, reporting on substantial accomplishments.

Elkader, Logan Co. Kansas
June 26, 1897

Dear Prof. Osborn:

We have put in a good week's work so far, and have had fairly
good luck. Have got Mr. Handel T. Martin's *Clidastes* [mosasaur]
skeleton up and ready for packing; have also partly worked out a
small turtle skull and limb-bones, and a gigantic turtle skeleton,
mostly complete but in very bad condition. It is in the Fort Pierre
beds in which fossils are quite rare and badly preserved, and was
largely exposed and broken up. With some careful treatment I believe it can be got to New York in such condition that it can be put
in order at the Museum. It is about six feet long, and may be *Pro-
tostega,* tho' we are not at all sure. We have had a Tylosaur [large
mosasaur] in sight which may turn out a good specimen, and a

number of smaller things. The weather has been fine except for a heavy thunderstorm. I hope to have a number of photos of characteristic exposures to show you when I get back.

Mr. Martin has engaged a man to work as teamster and cook and put in his spare time in collecting. He is paying him out of his own pocket at present, leaving the question open for next month as to whether the Museum can pay him. I doubt whether there will be enough left out of the hundred dollars I brought with me to pay this extra or even half of it, as I have had to get a barrel of plaster, and the wood for packing boxes will cost considerable—but I will see what can be done, if anything, towards it. If Martin pays his expenses the understanding is that Martin should keep anything that he (Baber) may collect. I think he has a fairly keen eye for collecting, tho' he has not done much work in that line.

In late July Matthew left Martin, who continued the work in the Niobrara Formation, and joined Jacob Wortman in Wyoming to collect Eocene mammals. But before leaving Kansas Matthew did some dickering with a couple of local collectors—the object being to obtain a magnificent mosasaur skeleton in their possession.

> Aurora, Wyoming
> July 25, 1897

Dear Prof. Osborn:

I left Elkader eight days ago and expect to be here a week longer. At Scott City saw Mr. Bourne's saurian. It is a Tylosaur skeleton, nearly complete and would make a fine slab specimen. It is very much crushed, the vertebrae are flattened out and the skull has been so far injured that the bones could not be replaced in their proper shape; they have had a slant-wise crush that has displaced them very much especially at the back. The nose is complete, very fine, the teeth as far as exposed are perfect. One paddle has been worked out; most, but not all of the bones are present though somewhat displaced. The other paddles may not be as good, but are apparently in fair condition. Some ribs are exposed in place but I do not know how many would prove to be missing when the specimen is worked up. It has been carefully got out, and I do not think much damage has been done to it. Bourne and his partner Morse (lawyers) own the specimen together. They have rather high ideals on the subject of fossils and would not sell for any such sum as twenty-five dollars I am pretty certain. Whether they would accept an offer of fifty I do not know; Martin believes that considerable of the bone material would need very careful treatment, as it is rather "punky." Dr. [Jacob] Wortman thinks from my description that the specimen might be worth a hundred dollars, as a maximum. I believe it would make a very fine slab, tho' more crushed than Williston's slab specimen at Lawrence. It is useless for exhibition in any other way. Under the circumstances I did not think it advisable to mention any sum to Bourne but got this much from him that he would rather keep the specimen and mount part of it in his collection than take a very small offer. He is an amateur, and has some fairly good things from the chalk. The arrangement was that the Museum would make him an offer based on my report.

I arranged with Baber on a basis of forty-five dollars for the month of July, for his services and such use as we ordinarily made of his extra team, and five dollars for two long trips, one to Oakley and one to Scott—fifty dollars altogether. He had not had much luck till the day before I left, when he found a shark skull which we hope may prove a fine one—it is the second specimen of the kind ever found in the Kansas Niobrara, so Martin tells me, the other having been found by Sternberg and sent to Zittel.[1] Martin has a fine saurian skull with atlas, axis and one cervical—it is about the size of our *Platecarpus,* complete and very little crushed.

Dr. Wortman is writing today about the work up here, so I will not say anything about it in this. I will be in N.Y. a fortnight from today. . . .

I told Baber that unless he had word from you to the contrary he would quit work at the end of this month. Should you wish to notify him there will be time for a telegram I believe, tho' not for a letter.

The sums initially offered to Bourne and Morse seem quite modest, but the final offer must have been satisfactory, because the specimen came to the American Museum, where today it can be seen—a truly magnificent exhibit, a complete skeleton almost thirty feet in length. In November 1898, when the skeleton was being prepared for exhibition, Will Matthew wrote a description of it to his father, including a rough but nonetheless striking little sketch.

Another fine skeleton we will have mounted by Christmas. It is a Mosasaur of the largest size and the most perfect one in many

[1] Charles Sternberg was a famous free-lance fossil collector, working in the late nineteenth and early twentieth centuries. Karl von Zittel was professor of paleontology at the University of Munich.

respects yet found. It has all the skull and body except the tip of the tail and will be mounted as a slab in the original chalk blocks in which it occurred, and the original position, the breaks between the blocks outside the specimen being filled in with plaster coloured to match the chalk. It shows a lot of cartilaginous structures unknown or very rare in any specimens hitherto found. Williston of Kansas University found this summer a much smaller species in which the paddle-webs and skin impressions are preserved, so that we are likely to get a great increase in our knowledge of these Sea Lizards, as to what they really looked like both inside and outside. Our specimen is twenty-eight feet long when straightened out, with about two or three feet of tail fragmentary or missing. It lies in the rock like this. [Sketch of specimen.]

The lower jaw is 46 inches long. (letter of November 17, 1898)

This letter was written after Matthew's return from his second western field trip, a collecting season of particular significance to him, because it was the first expedition entirely devoted to his own particular interests. The previous summer had been a breaking-in period, first with Handel Martin in the Kansas chalk and then with Wortman in the Eocene beds of Wyoming. But the summer of 1898 was all his own, and he made good use of the opportunity. This was the year he made large and significant collections in northeastern Colorado of Miocene mammals from stratigraphic levels hitherto unknown. Those were the collections upon which he based his monograph entitled "Fossil Mammals of the Tertiary of Northeastern Colorado," published in 1901 as a quarto-size Memoir of the American Museum of Natural History.

Unlike the previous year he got an early start, so that by early June he was in the field. It was a cold soggy time, so unusually damp that "the oldest settler doesn't remember such weather," Will wrote his mother. But then it turned hot with a vengeance.

Sterling, Colorado
June 27, 1898

My Dear Mother:

I am stewing in a tent just at present but not lying on my back in a crack in the rocks, and have no intention of taking the latter position unless it affords due shelter from the violent sun. It is hot here in the day time but quite cold at night and the weather has settled fine. We have explored the Kansas-Nebraska boundary without success and are now going up against the monsters of Colorado, where I hope we will do better. I have just heard from Dr. Wortman what he is unable to leave the work in Wyoming so I must remain in charge of this party, and will not get back to N.Y. till near the end of July. This climate suits me very well, and I am gathering a heavy coat of tan and a wire-nail digestion. The days are at their longest, and a good day's work leaves one just about ready for bed after supper. It would be more pleasant if the rest of the party were a little better companions; but of course one can't expect everything, and though their language and ideas are not exactly Emerson's we get along very well. We have had very poor luck so far but hope to do better in this field. . . .

He then goes on to describe the Colorado countryside (east of the Rockies) to his mother: the cacti and the flowering yucca, other wild flowers, and the fields of alfalfa in the river bottoms. And great fields of grain "sometimes stretching a mile square or more, and looking very fine this year of rain."

Just how productive his summer had been is revealed in his letter of November 17, 1898, to his father.

Our Colorado collections are panning out even better than we expected. Besides our huge giraffe-camel we have fine material of various intermediate forms which will make the camel series the most complete known. And some of the specimens are so perfectly preserved that they are indistinguishable except by weight from recent bones bleached for exhibition. I have never seen any quite so perfect as this. We have material enough in this summer's collections to keep us busy for several years preparing it for exhibition, and to half fill a hall as large as our present one with these collections alone.

Will Matthew's enthusiasm concerning the fossil bones he had collected in Colorado will strike a harmonic note in the heart of anyone who has collected such fossils in the High Plains. As he remarked, at first glance one would not take them for stony objects, so perfectly do they resemble modern bones. The form and texture of these fossils are such as to give to them a truly esthetic appeal—of the kind that led the great American artist Georgia O'Keeffe to introduce into her luminescent paintings the skulls, vertebrae, or pelvic girdles of deer or cattle that she had picked up in the badlands near her desert home in New Mexico.

While Matthew was praising his Colorado fossils as esthetic objects he was also quite cognizant that, as a result of one summer's work in the field, he and his associates had accumulated enough material to "keep us busy for several years." He mentions preparation (cleaning, mending, etc.) and exhibition, in his description of "keeping busy," but more important was the research on these fossils, the descriptions and analyses, that would take much of Matthew's time and effort between the autumn of 1898 and the day in 1901 when his Memoir was published.

Because of this rich collection, offering as it did many avenues of research to follow, Matthew apparently did not feel pressured to immediately return to the field in search of more fossil *mammals,* and so it was that in the summer of 1899 he participated in a huge "dig" in Wyoming, where gigantic dinosaurs were being excavated.

Two summers earlier, while Will worked in the Kansas chalk beds, an American Museum field party composed of Jacob Wortman, Walter Granger, Barnum Brown, and H. W. Menke, a field assistant, had explored the region around Como Bluff, Wyoming, where—twenty years earlier—O. C. Marsh's collectors had unearthed the bones of late Jurassic dinosaur giants. Osborn wanted some huge fossil dinosaurs on display in New York. Among other considerations, giant dinosaurs in the museum halls would attract the public and bring strong support to the institution. So the American Museum search began. The next summer, while Matthew was finding the beautiful Tertiary mammals in northeastern Colorado, the dinosaur party discovered a rich site a few miles to the north of Como Bluff. This locale was christened "Bone Cabin" because a sheepherder had built a cabin using large fragments of the dinosaur bones profusely scattered over the ground.

The bone diggers began their work, and at the end of that field season thirty *tons* of dinosaur bones were shipped back to New York. (Work continued through the following five years, with the result that more than fifty additional tons of fossils went to the American Museum.)

In July 1899 Matthew wrote to Osborn from Medicine Bow, Wyoming, reporting their progress. The party had opened a new quarry four miles from Bone Cabin, the "Nine Mile Quarry" from which a brontosaur was obtained. It should be noted that several quarries were opened and excavated in this general region.

Medicine Bow, Wyoming
July 2, 1899

My Dear Prof. Osborn:

I arrived here about ten days ago, and have been mostly at the new quarry since. This new quarry is on the Little Medicine four miles below Bone Cabin Quarry, about half a mile from the Nine-Mile Crossing (of the Little Medicine). The specimen is a Bronto-saur, and corresponds closely with the one from Aurora got in 1897. Parts preserved are all of a single individual. The best part is a connected series of four cervical and nine dorsal vertebrae with the left ribs in place and all or nearly all the right ribs displaced. There are also about a dozen caudals, one cervical, both coracoids and pubes, femur, parts of sacrum and ilia and various fragments of vertebrae ribs and limb bones exposed. There may be considerably more of the animal not yet exposed, but do not look to find any connected series of vertebrae or complete limbs. One tooth is the only recognized fragment from the skull as yet. Carnivore *teeth* occur with the animal, and the floor on which it rests is littered with small unrecognizable fragments of bone. I suspect that in this and in most cases, the fragmentary condition of the skeleton is due not to currents or weathering, but to the carnivorous dinosaurs which in this case seem to have left the bulky body intact but to have torn up and destroyed the extremities. The ilium and sacrum are broken up and other limb bones damaged while the delicate vertebrae and ribs are perfect. The bone is very fine and the matrix excellent. Ends of limb-bones and other cartilage-covered surfaces a little soft, but all other parts extremely hard, black and un-crushed. This specimen assures us of the back and ribs for the mounted skeleton, and we may get the limbs under the bank, as I have seen no fragments of them. Cervicals I am afraid we shall have to reconstruct if at all from fragments of which I have already taken up a number. Sacrum may be reconstructible if we get more pieces of it. We are saving all recognizable fragments of importance as they may be unable to reconstruct some of the bones.

Dr. Wortman was here a few days ago. He has been in the Freeze-out Mountains about fifteen miles north, but has had poor luck and is now near the mouth of the Little Medicine, where he hopes to get better specimens. Granger rather suspects that he came up to investigate this same specimen that we are working on; it was one which the party found last year, and Granger had told Wortman some time ago that he did not know whether it was worth investigating—it had a very unpromising outcrop. So the Doctor may have come up to see what there was in it; but he did not let out anything of the sort if it was so.

Prof. [Richard S.] Lull makes a first-class man in camp, and does good and careful work. He is a very nice fellow personally. I think he would like if possible to get some prospecting experience, which would of course be more instructive to him; also he has some plans for trying to do some post-grad. work on vert. paleaont., which he hopes to talk over with you when you come out.

Kaison [sic] is an excellent workman and the most industrious fellow I have seen. Would it be possible to get him in the Department this winter? He will not stay out here another year, and it would be a great pity to lose him or have him snapped up by some other museum.

Spite of the Doctor's ill success, [Walter] Granger and I are going to do a little prospecting in the Freeze-out. Granger had a couple of prospects that he wants to look up, and there is a good deal of country to explore. There is considerable Tertiary north and west of here, which we won't especially go into just yet, though some of it may be worth exploring.

Prof. Knight was ill with measles, when I stopped at Laramie, and could not see anyone. No-one seems to know much about his great aggregation, as to where they will go or what they will do. Dr. Wortman thought they would go into the Big Horn, find next to nothing, and return disgusted. I hope it may be so.

All well, and preparing to celebrate the Fourth with beer and skittles (whatever they may be). I expected we'll know all about our Brontosaur about the time you come out as it is heavy work to get it out from under ten feet of rock, as we now have at back of quarry.

A picture taken that summer shows the field party of 1899: Osborn (who had come out on an inspection trip), Matthew, Granger, Lull (eventually a professor at Yale), Thomson and Kaisen (preparators at the American Museum), and a field assistant named Snyder. For one interested in the history of paleontology it is an informative picture—a record of some young paleontologists who, during the first three decades of the new century, were to become leading figures in their science. Osborn, characteristically clad in a jacket and wearing a necktie, is obviously a visitor—much too neat to have been digging fossils. Matthew, having rid himself of the mustache and beard he had been wearing a year or two previously, looks remarkably boyish. He—like Granger, Lull, Kaisen, and Thomson—bears the stamp of the bone-digger: rough clothes and informal headgear.

Perhaps this picture might be regarded as epochal, as well as informative. Not only does it record the excavation of dinosaurian giants during the final year of a century in the early years of which dinosaurs were quite unknown, but it also marks the beginning of a new paleontological era in North America. On March 18, 1899, Marsh had died in New Haven, and with his passing, just two years after the death of Cope, the pioneer period of American vertebrate paleontology had come to an end. Changes were bound to occur.

On March 24 Will Matthew had written his father "You will have heard of Prof. Marsh's death. He has left all his money to Yale, for scientific purposes. . . . The general opinion is that Dr. Wortman will be appointed his successor. There seems no other man of equal claims. . . . It will be a new era in New Haven if Wortman takes charge; he will publish and illustrate a vast mass of fine material that has never been touched."

But it didn't work out that way. Wortman resigned (shortly after Matthew wrote that letter) and went to the newly established Carnegie Museum in Pittsburgh, and Matthew was advanced to associate curator, as Wortman's successor in New York.

Thus at the end of the nineteenth century Will Matthew was solidly established as a prominent younger member of his profession, actively engaged in paleontological research and fieldwork, participating in various ways in the museum's programs, and looking forward to what the years ahead might bring. As for the immediate future, there was his first trip to Europe, where he would become acquainted at first hand with paleontologists and museums on the other side of the Atlantic. This trip would occupy his summer of 1900.

CHAPTER 7

Europe

*I*n the spring of 1899, when Will Matthew was making plans for a summer of fieldwork at the Bone Cabin dinosaur quarries, he was also looking ahead to the summer of 1900, when there was to be not only a World's Fair but also the quadrennial International Geological Congress in Paris. He dearly wished to go abroad the next year, to see the collections and visit some of the places where the science of vertebrate paleontology had its beginnings. And he wished to meet his European paleontological colleagues, with many of whom he had corresponded. The meetings in Paris, a magnet for people of his persuasion, afforded an opportunity not to be missed. But could he obtain funds for such a trip?

He was certainly feeling pinched, particularly because he was still shouldering much of the responsibility for George.

> 509 Amst. Ave.
> April 10, 1899
>
> *My dear Father:*
>
> I have delayed answering you in regard to what I could do towards George's education next year, as I had to see Prof. Osborn about my own prospects first, and had no opportunity till today.
>
> I will be glad to do what I can towards helping George to take

a post-grad. course, as I do not think the money would be wasted. He is making a good record this year, and is likely to go on doing so. But I have been sailing too close to the wind the last four years and have been barely able to make both ends meet. . . .

I wish I could promise more, but I have been unable to meet necessary expenses this year . . . and also I find myself unable to meet social duties, or to join any scientific societies on account of the expense involved; a condition of things which I ought not to allow myself to become involved in.

As to the Paris and Continental trip, that is more or less of a day-dream at present. I *may* be able to carry it through, and it will do no harm to make preparations. But it is quite likely I may not be able to 'raise the wind.' It is very kind of you to offer to help me out, and I appreciate your offer very much, but I will not go unless I have the money saved up.

Could he have possibly saved sufficient funds from his own meager salary, subject as it was to the drain imposed by George's continuing studies at Columbia? Hardly, one would think. It seems probable that Osborn secured the necessary wherewithall for the trip—Osborn came to the help of his departmental associates on many occasions. However that may be, Matthew obviously had his European trip planned and secured in the spring of 1900. In May he sailed on the *Maasdam* bound for Rotterdam, where he landed on May 30.

It is probable that few people under the age of fifty in this era of air travel can appreciate the experience of a transatlantic voyage. In the days of ocean liners, especially at the turn of the century, a trip from New York to a European port was a venture not lightly undertaken. It was a journey far different from the casual five- or ten-hour flight of today.

There were numerous problems of planning and packing, for one had to carry clothes and other supplies for an extended trip, and thus required rather cumbersome luggage. There were the trip to the pier accompanied by friends and relatives, the lengthy farewells, and the frenzied process of getting aboard and established in one's quarters. Then came the journey, a matter of eight to ten days when one lived a different life, far removed in spirit and substance from life on land. In many respects this enforced vacation was a blessing; it afforded the traveler a respite, a time to sever ties with the land disappearing over the horizon to stern, and a time to prepare for the adventures ahead. This was one of the great advantages of a steamship voyage.

But crossing the ocean was not always a joyful and carefree event, because all too often there was that horrendous disorder, mal de mer, to make life miserable. (In the old days of sailing ships—when sometimes an Atlantic crossing was six weeks of torture in stormy seas—it was not unheard of for passengers to die of seasickness.)

Eventually the port of destination was reached and after the frenetic ordeals of disembarkation and customs were endured, one was safely on dry land trying to adjust to stable footing after a heaving deck. Is it any wonder that the land across the water seemed more removed from home than is the case today?

Little information remains about Will Matthew's days aboard ship. Evidently, the time at sea passed in the usual way, with alternating periods of eating, sleeping, reading, visiting, enjoying amateur entertainments, and enduring periods of boredom. Excerpts from an all too scant journal:

> Concerts Monday and Tuesday—local talent; some very remarkable—Palmer, who gets them up is really a very able sort of chap for such—speaks offhand well enough—tho' not a good English speaker. Frenchman with an English name—teaches French in Boston. Younger brother also language teacher. Sanford—Tufts man—philosophy his ideal, philology his actual—good fencer and athlete—teaches at Naugatuck, Conn. Porter—medico—Beta—probably *not* athlete. Kimball—artist—very seasick at first—pleasant sort, as is probably also *Howland*—skull cap—whistled solo for concert—genial remarks.

With these traveling companions Will Matthew whiled away the hours when he wasn't reading or observing the passing scene. He complained in the journal of a rainy spell lasting for three days when the sheltered parts of the decks were too crowded with uncomfortable passengers. But then the sun came out and passing ships were seen—two of them with sails. "One rather near, 3 masts, mizzen not square rigged" wrote the young man familiar with these lingering survivors from the Age of Sail. And he saw great numbers of Portuguese men-of-war, and was on the lookout for spouting whales.

An overheard remark was noted: "I am ze Ingleesh interpretaire,

I spik ze languish wis correctness and fluencie. Rosbif-goddam." And some Matthew verse appears:

> Some try for a character free
> from all spots
> But 'tis my aim in life to
> accumulate blots.
>
> A mute inglorious Milton
> I contain
> Great thoughts nipped in
> the bud.
>
> A little picnic on the
> Maasdam's bow
> A pack of cards, a modicum
> of pop
> A jest, a laugh, a story or
> a song
> 'Twas near enough to Paradise
> I trow.
>
> 'Tis a voice from the smoking-
> room, heard to complain
> "They have turned us all out
> 'leven thirty again
> Come up and we'll walk on
> the deck in the rain."
>
> A rolling deck, from a
> seasick wreck
> Gets plenty of strong abuse.
> But when walking there
> with a lady fair
> You'll find that it has its
> use.

On May 30 he came ashore in Rotterdam, to begin three months of travel, interspersed with scientific sessions in Paris and study sessions at various museums. His itinerary included France, Belgium, Holland, Germany, Switzerland, and England, the sort of schedule that might have been followed by almost any tourist in those days, except that Will Matthew was not a tourist in the ordinary sense. He was there for serious purposes; by the end of three months he had filled a little notebook with observations and drawings of fossil mammals seen in museums.

Yet even the most dedicated paleontologist cannot ignore the local scene, and Will managed to take some holidays from fossils to see the sights. He met his cousin Christina Matthew (the daughter of Robert and Christiana, his father's brother and sister-in-law), and together they had some walking tours in the Alps. Then came September; the meetings in Paris were over, and it was time to go home.

The journey back to New York, on the *Potsdam* of the Holland-America line, was tedious (a twelve-day crossing), stormy, and altogether not very pleasant.

> I haven't found it easy to do anything to pass away the time except read, and I've read till my eyes are sore and my glasses broken. Contrived to fix up the latter roughly or I'd have been in a bad way. Haven't been at all sea-sick—this weather is fine to see, for this little boat only 350 ft. long is tossed about in great shape in the stormy weather we have been having. Much to the destruction of crockery etc., which one hears smashing around very extensively.
>
> It isn't possible to write a very respectable hand in this jar and roll, so you must excuse the illegibility hereof.
>
> (letter to mother, September 19, 1900)

In New York he settled down in a new lodging on West 145th Street, between Broadway and Amsterdam Avenue, this after several years at 509 Amsterdam Avenue. And soon after moving into his new apartment he wrote to his father.

> 407 W. 145th Str.
> Oct. 6, 1900

My dear Father:

> I'm settled down in the new place now, and think E. [his sister Elsie, there to keep house for him] and I are going to be very comfortable here. . . .
>
> I spent about a month in Paris, chiefly in attending meetings and excursions and studying the collection at the Jardin des Plantes. Prof. Barrois assigned me the honor of reading your paper before the Congress—and I found that your name was considerable of a passport in the way of introductions. As you know I didn't see very

many of the English and French cities—only London in England, and was fully occupied there in making studies and notes at the Natural History Museum.... There were not very many Americans at the Congress, and very few English,—mostly French and Germans with a considerable representation of other continental nations.

The rest of the letter was devoted to details on collections he had studied and fossil localities he had visited.

CHAPTER 8

Kate Lee

Kate Lee was born on April 2, 1876, in Troy, Pennsylvania. Her parents were James Edgar Lee, an ex-Confederate soldier and Jane Eliza Pomeroy, both originally from Montgomery, Alabama. By the time of Kate's birth they had moved to Pennsylvania because they had found the problems of living in the South after the Civil War too difficult. James Lee had been a volunteer member of the "Montgomery Blues," one of the numerous military companies that had sprung up in anticipation of the coming conflict.

In Troy Mr. Lee was the proprietor of a store that sold groceries as well as good china. There he prospered in a modest manner for years, and the family grew until there were eight children—Charles, Bess, Emma, Edward, Kate, Mary (Mame), Montague (Mont), and Pomeroy (Roy).

When Kate was six years old she was struck by an affliction that profoundly changed life in the Lee household. Something affected her eyes—evidently a corneal infection—and she became functionally blind. For nine years she was confined to a dark room; when she did venture out it was with heavy bandages over her eyes.

It was a tragic situation but fortunately she had a loving family and everything that could be done to help her through those years, was done. Her parents and siblings talked and played with her in her room, and when she came out with the bandages over her eyes they would read to her by the hour. Her brothers and sisters regularly shared their homework with her, often sitting outside her door and reading their assignments to Kate—there in the dark on the other side—and discussing the lessons with her.

Consequently, even though she had no formal schooling during those nine years, she did have the benefit of informal instruction, passed on to her through a door that was simultaneously a portal and a barrier between her two worlds. By means of such shared lessons and readings Kate Lee acquired an unusual exposure to literature, with the result that, while still young, she had become familiar with the plays of Shakespeare, the novels of Dickens (which she particularly enjoyed), and other literary works of the same caliber, to a degree remarkable for her age. Before the loss of her eyesight, she had learned to read at a fairly high level for a young child; therefore she had a real appreciation of the works she listened to during the family reading sessions.

One interesting detail: In later years she told of how every evening when she was little her father would bring home a handful of pennies from the store, open the door to her dark room, and fling the pennies in. She would then feel about the room, picking up pennies as she found them, thus adding a little each day to her tiny fortune, while gaining confidence in utilizing senses other than sight.

When Kate was fifteen the Lees decided that this state of affairs could not continue; they moved to Brooklyn where there was a doctor who, friends had informed them, might be able to cure Kate's affliction. He was.

But thereafter Kate was extremely short-sighted and always had to wear thick glasses. Nevertheless, she was elated to have her sight restored; for the rest of her life she had a keen visual appreciation of the world around her. To her the world was, all in all, a beautiful place; a field of flowers, a sunset, brought tears to her eyes.

Naturally Kate had a lot of catching-up to do after her nine years of blindness, and it would appear that she managed to overcome satisfactorily any deficiencies resulting from her afflicted years. She soon became adjusted to a world of light and color.

This world was centered in Brooklyn, where the Lee family had come to reside. Although Brooklyn today may not call up visions of

glamour, in the final decade of the nineteenth century it was for Kate
Lee a city of considerable enchantment. In those days she made occasional diary entries that provide glimpses of life in the Lee household,
and in the semi-rural borough across the East River from Manhattan.

There was not much money, but there was an abundance of family
love and activity. With a tribe of attractive young people in residence
visitors were numerous. One gets the impression that the rooms were
filled with talk and laughter. Among the frequent visitors were
nearby neighbors—especially the Dillers and the Blackmans.

Kate's life was a constant succession of visits (at home and elsewhere), of little trips around Brooklyn and Long Island, and on occasion to Manhattan, of sewing and reading, of music at home, of
theater and concerts (in Brooklyn), and of family celebrations. It was
a gay life, but there were serious times, too.

She went to school; on January 5, 1892, she wrote in her diary of
beginning school after the Christmas vacation, and on February 6th
she noted with pride that she had been promoted to the sixth grade—
this at the age of fifteen. She was making up for lost time, but her
schooling was (as she later recounted) to be limited to two years of
formal instruction. Furthermore, her studies, especially in the first
year after she had regained her sight, was interrupted at times by
lingering problems with her eyes. On March 12, 1892, she wrote that
her "eyes have kept me home from school for three weeks, but I have
kept up with my studies as far as I could."

So the months passed, and she devoted herself to such schooling as
could be managed, but largely to the activities of a young lady living
at home and helping with the domestic labors necessary in a house
inhabited by a large family.

Of her siblings Kate was especially attached to two sisters, Bess—
eight years her senior, and Emma—two years older than she. Despite
the age differences they engaged in many activities together.

Some entries in her diaries of 1892–93 tell a little story with an
almost Dickensian flavor, a story revealing certain social conditions in
Brooklyn in days before any well-organized welfare programs existed.
A little boy named Hyram came to the Lee house selling paper
flowers:

> Little Hyram was here last night and a more peculiar little baggage
> I never saw in my life. He used to come around some time ago

with paper flowers to sell. Mama told him to come after Xmas, and he might have the boys' coats.

When he came in he wished us all a Merry Christmas and Bess asked him what kind of a Christmas he had. "O!" he said, "as well as we could afford, Miss." Bess said, "Well, what did you get?" "Oh, nothing at all," in a very cheerful little voice. Mame fixed a place for him at the table and Mama gave him some turkey and potatoes and tomatoes and things to eat. After a while I heard Mama say "Don't you like turkey, Hyram?" "Well, yes," he said, "but my mother didn't have any Christmas dinner, so if you don't mind, I'll just take this home to her." So we gave him some more to take to his mother and he pitched into what he had with a lively appetite.

A few days later:

Hyram came down in the morning, and Bess and I went home with him. He is just as honest as the day is long, and very cheerful, too. We took the Third Ave. car up to 42nd Street, and he took us down toward the shore, through mud up to your knees (slightly exaggerated) and up and down over sandy humps until we came to a little unpainted house, where he took us in. When we got inside, which we did through a door whose sill [sic] hit me in the forehead, we found them, viz. the mother, father, sister and brother all seated at a table covered with white oil cloth. The father, George Hymas, is a painter by trade, but has lost employment through his wrists being paralyzed, having set in from lead poisoning.

Hyram gets three dollars a week during the holidays at a toy store, but of course that employment ceases now.

We got Hyram to go with us up to the bakers where Bess got three loaves of bread and also to the grocers where she got 3½ pounds of sugar, 1 pound of cookies, 1 quart of apples, one can of tomatoes and one can of condensed milk. She wanted to get some meat, but neither of us knew what kind to get. So Bess gave him fifty cents for his mother to get it with and after we had told him to wish her a Happy New Year he said, "Well, I must shake hands with you," which he did with his little brown fin and a beaming countenance.

Of all the household activities perhaps none afforded Kate more pleasure than sewing. She seemed always to be working on something with needle and thread—dresses and other articles of clothing, not

1. Portrait of William Diller Matthew
by Peter VanValkenburgh, 1929.

2. The Matthew family of Saint John, New Brunswick, 1871. The men are (*left to right*) Charles, Robert, George Frederic, and Douglas. In the front row (*left to right*) are Bessie Matthew Tilley, sister of the four Matthew brothers; Katherine Diller Matthew, George's wife; and Christiana Diller Matthew, Robert's wife.

3. The George Frederic Matthew home, "Hillside," in Saint John, New Brunswick. It was here that Will Matthew spent his boyhood years, after the great Saint John fire of 1877.

(a)

(b)

4. A reconstruction of the discovery of the giant trilobite, *Paradoxides regina:* (a) George Frederic Matthew sets out with his son Will for some fossil hunting in the outskirts of Saint John; (b) Will discovers the fossil, much to the delight of the two fossil hunters; (c) The rock containing the fossil has been excavated, and is brought home and admired by fa-

(c)

(d)

ther and son; (d) George Frederic Matthew writes a description of the fossil, published some years later in the *Transactions of the Royal Society of Canada*. He proudly named the fossil *regina*, in honor of Queen Victoria. Drawings by Margaret Matthew Colbert.

Gondola Point N. B.

5. The Kennebecasis River at Gondola Point, north of Saint John. The Kennebecasis here is a broad estuary. The house in the center is "Amkuk," the summer residence of the George Frederic Matthew family. At the right is seen a steamer, the *Hampton,* which in earlier days had plied the Kennebecasis. The steamer is docked at Fluellings Wharf, the scene of the unfortunate dunking of Kate Lee Matthew, when the canoe capsized with all the gear for the honeymoon camping trip (see chapter 9).

7. The Columbia School of Mines on Forty-ninth Street, before it moved uptown to Morningside Heights.

6. William Diller Matthew
as a student at the
University of New Brunswick, 1889.

Acc't of expenses Jan 1 - 19

Jan'y 1	Rec'd from Father		9.00
" 4	B'k'fast on State of Maine	.50	
" "	Car Fares. Boston	.30	
" "	Parcels	.10	
" "	Tea	.25	
" "	Fall River line ticket	3.00	
" 5	Car Fares N.Y. & Bklyn	.10	
" 6	Tickets	1.00	
" 6	Church	.10	
" 7	Tickets	.25	
" "	Trunk moved	.75	
" 8	Letter Paper	.30	
" 10	Lunch	.50	
" 11	Rec'd. fr. Mr. Pike		30.00
" "	Bag	2.00	
" "	Lunch - Box	.40	
" "	Board	6.00	
" "	Laundry	.79	
" 12	Church	.10	
" 13	Y. M. C. A. key	.50	
" 13	Laboratory keys	.75	
" "	Newspaper	.03	
" "	Matches	.05	
" 15	Lunch	.50	
" "	1 ft Plat. wire .40 Matrasses .70	1.10	
" "	Locker key	.75	
" 17	New Apron	1.25	
" 18	Qual. Lab. Key	.25	
" "	Laundry	.65	
" 18	Board	6.00	
" 19	Church	.10	
" "	Bal'ce to date	10.53	
		39.00	39.00

W. D. Matthew

8. Will Matthew's expense account for the first nineteen days of 1890.

9. James Furman Kemp, the beloved "Uncle Jimmy" who for many years guided the Department of Geology at Columbia University. Kemp and Will Matthew had a special relationship, because Will Matthew was one of Kemp's first graduate students.

10. Henry Fairfield Osborn in 1890, just before he came to New York from Princeton. In New York he founded the Department of Mammalian Paleontology (later the Department of Vertebrate Paleontology) at the American Museum of Natural History and simultaneously was a professor in the Department of Biology at Columbia College. (Osborn was also dean of the graduate faculty at Columbia.) Here we see the very proper young patrician at the beginning of his distinguished scientific career, which was paralleled by his social status in New York society.

PHYTOSAURS

11. *Rutiodon carolinensis,* the fossil reptile that was excavated by Will Matthew in 1894 and 1895 in a hot, muggy coal mine near Egypt, North Carolina. His labors in the mine (see chapter 5) constituted his first fieldwork for the museum. *Rutiodon* was a phytosaur that dominated the scene when the first mammals inhabited North America. The phytosaurs were crocodilelike reptiles that preceded the crocodiles but were not their ancestors.

12. The American Museum of Natural History, as it existed when Will Matthew began his paleontological career there in 1895. On the left the original building—Section 1, or the "Bickmore Section"—is partially visible. In the center is Section 2, the "Memorial Section," the building that for many years served as the main entrance to the museum. On the right is Section 3, the "North American Section," facing Seventy-seventh Street.

13. Will Matthew in his bachelor quarters at 509 Amsterdam Avenue, between Eighty-fourth and Eighty-fifth streets, New York.

14. Will's sister, Bess Matthew, who, before her marriage to Edward (Ned) Manning, took turns with her older sister, Elsie, in keeping house for Will (and such brothers as might be living with Will) in Manhattan. Bess Matthew was a talented musician, as was her husband.

15. The twin town houses at 2102 Pine Street, Philadelphia, where Edward Drinker Cope spent the last years of his life. Here he had his office, his laboratory, his storage rooms, and, in one back corner, a cot, which constituted his bedroom.

16. The magnificent rat's nest that was Cope's study. The objects and publications littering Cope's desk and all other horizontal surfaces in the room are indicative of his broad interests in vertebrate anatomy and evolution. Fastened to the wall next to the door is what appears to be a dinosaur leg bone. Charles Knight, the famous artist who spent a lifetime depicting extinct animals on canvas and paper, once told me that such clutter was pervasive throughout the Pine Street houses. Even the stairways were filled with specimen bottles and other impedimenta, so that one had to pick one's way carefully.

17. Cope in his office-study at the Pine Street house. The tense pose and penetrating stare of this scientific genius are characteristic of a man whose inner demon drove him to incredible accomplishments in the disciplines of ichthyology, herpetology, mammalogy, and vertebrate paleontology.

Lake Basins   F 13-27 (w)

Period	Epoch		Basin	Thickness	Bed	Fauna
	Pliocene	Equid. Plist. Rec.	Equus & Megalonyx	100		Elephants & Mastodons, Horses & Camels come other animals
	Pliocene		Blanco	100 or 150		Mastodons & Elephants, Horses and Camels, Bear, Cats, Dogs etc
			Palo Duro			
	Miocene		Loup Fork	400		Mastodons Rhinoceroses, Oreodonts, Camels, Horses, Deer, Cats, Dogs
	Miocene		Deep River	300		Rhinoceroses Oreodonts, Camels, Horses, Deer, Cats, Dogs (Elephants? Mastodon)
	Miocene		(Interval not represented in America)			
	Miocene		John Day	1000		Rhinoceroses. Oreodonts, Horses Camels, Tapirs, Primitive Deer, True Carnivores (Cats Dogs)
Tertiary	Oligocene		White River	800 to 1000	Protoceras Bed	Rhinoceroses. Anthracotheres, Oreodonts Tapirs Horses, Camels, Primitive Deer, True Carnivores
Tertiary	Oligocene		White River		Oreodon Bed	Giant Pigs, Rhinoceroses, Oreodonts & true Carnivores, Tapirs, Horses Oreodonts, Camels, Primitive Deer
Tertiary	Oligocene		White River		Titanotherium Bed	Titanotheres, Giant Pigs. Rhinoceroses, Oreodonts & True Carnivores, Tapirs Horses Oreodonts, Camels
Tertiary	Eocene Upper		Uinta	600	Diplacodon Bed	Titanotheres, Giant Pigs, Rhinoceroses. Oreodonts
Tertiary	Eocene Upper		Uinta		Telmatotherium Bed	Tapirs. Horses. Oreodonts, Camels. Primates
Tertiary	Eocene Middle		Bridger	2000	Uintatherium Bed	Amblypods, Titanotheres, Rhinoceroses. Horses, (small) Giant Pigs (Elotheres), Creodonts Primates Edentates Artiodactyls
Tertiary	Eocene Lower		Wind River	800	Bathyopsis Bed	Amblypods, Titanotheres, Horses. Tapirs. Creodonts, ? Rhinoceroses, Artiodactyls. Primates. Edentates
Tertiary	Eocene Lower		Wasatch	2000	Coryphodon Bed	Amblypods, Horses. Tapirs, Creodonts, Edentates
Tertiary	Eocene Lower		Wasatch		Meniscotherium Bed	Condylarths, Primates. Artiodactyl
Tertiary	Eocene Basal		Torrejon	500	Pantolambda Bed	Monotremes (allies to), Condylarths, Creodonts, Edentates, Amblypods, ? Primates
Tertiary	Eocene Basal		Puerco	500	Polymastodon Bed	Monotremes (allies to), Condylarths, Creodonts, Edentates
						Est'd age 2,900,000 (Walcott)
Cretaceous (?)			Laramie	1000 to 5000		

18. A chart of the freshwater Tertiary sediments of western North America, included by Will Matthew in his letter to his father of March 10, 1896. This compilation is interesting as the forerunner of the diagrams and lists by Matthew in his paper of 1899, "A Provisional Classification of the Fresh-Water Tertiary of the West."

19. Will Matthew (*center*) and his cousins Angela and Alfred Diller on a rustic outing. Angela Diller later founded the Diller-Quaile School of Music in New York.

20. Will Matthew, in the fancy socks, on a bicycle outing with two friends. The locale appears to be suburban New York—perhaps Westchester.

21. Othniel Charles Marsh of Yale University, Cope's bitter rival. The cold, scientifically ruthless nature of Marsh can be seen in this portrait. (He was often referred to by various contemporaries as the "Great Dismal Swamp.")

22. The staff of the Department of Vertebrate Paleontology at the American Museum of Natural History, in the late 1890s. *Front row, left to right:* Adam Hermann, chief of the preparation laboratory; Jacob Wortman, curator; William Diller Matthew, associate curator; Walter Granger, assistant curator. The scientific preparators in the back row are (*left to right*): Charles Thompson, Alexander Edgar, Charles Christman, Thomas Carr, Frederic Schneider, and Arthur Coggeshall. (Wortman, a medical doctor by training, had been Cope's assistant before coming to New York.)

23. The Osborn Library at the American Museum of Natural History in 1905. The delightful muddle of skulls, jaws, feet, and other bones shows that this room evidently was being used as a study area as well as a library. The fossils are perissodactyls, or odd-toed hoofed mammals (represented today by horses, rhinoceroses, and tapirs), and their abundance illustrates the richness of American Museum collections, even at an early date in the history of the institution.

24. The American Museum camp at the Bone Cabin Quarry, 1899.

26. Early morning at the Bone Cabin Quarry camp. Osborn is awake and sitting up. Matthew is coming around, leaning on one elbow, while another bone hunter (perhaps Lull?) is still comatose, wrapped in his tarpaulin. In the background Pete Kaisen is cooking breakfast.

25. Fossil collectors at the Bone Cabin Quarry camp. *Seated, left to right:* Walter Granger, Henry Fairfield Osborn, William Diller Matthew. *Standing, left to right:* Snyder (a temporary summer assistant—perhaps the camp cook—whose first name is not in the records), Richard Swann Lull, Albert Thomson, and Peter Kaisen. The dog's name was Jack.

27. Getting down to the bone layer at "Nine Mile Quarry" at the Bone Cabin project. *Left to right:* Kaisen, Lull, and Matthew. Note Matthew's improvised havelock, to protect his neck against sunburn; for him the sun was an implacable enemy. In the foreground, some partially plastered bones.

28. A treasure excavated from the Bone Cabin Quarry—a gigantic brontosaur. Here the skeleton is seen as it was displayed at the American Museum for many years—a mount that was originally supervised by Matthew and Osborn. This is a mount of the original fossil bones. Missing parts have been restored and given a slightly different color to distinguish them. Matthew was a strong proponent of exhibiting original fossils.

29. Kate Lee, age eighteen (*left*), and her sisters Emma (*center*), and Bess (*right*), 1894.

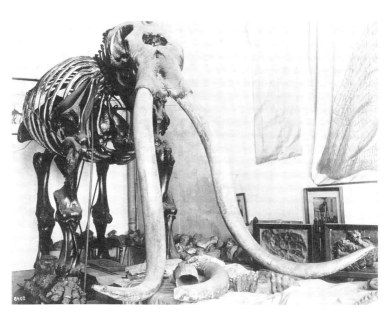

30. The Warren Mastodon, as it was mounted in Boston. The papier mâché tusks are fancifully elongated to ridiculous lengths. (Parts of the real tusks are seen on the floor between the artificial tusks.) The body is set too high and laterally expanded. The bones are covered with a thick, dark varnish. It was Will Matthew's task to supervise the disassembly and packing of this skeleton for shipment to New York.

31. The Warren Mastodon remounted at the American Museum of Natural History. Here one can see the beauty of a fossil skeleton properly posed and assembled.

32. Matthew's camp on Porcupine Creek, South Dakota, in 1906. The sheet iron stove with its long stovepipe along with the wall tent, chuck wagon, and horses reveal a past perhaps forgotten or unappreciated by many present-day fossil hunters.

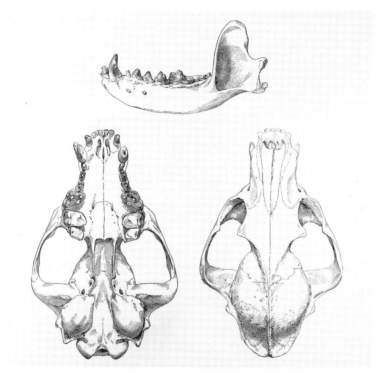

33. *Phlaocyon leucosteus,* an ancestral raccoon described in Matthew's 1901 monograph on the fossil mammals of the Tertiary of northeastern Colorado. This is one of the specimens that inspired Matthew's enthusiastic comments about the state of their preservation. The drawing is an example of the superb scientific illustrations done at the museum under Matthew's close supervision, especially by Mrs. Lindsay Morris Sterling and Erwin Christman.

34. The Matthew house at Hastings-on-Hudson soon after completion. The architect was William Sanger.

35. The former Matthew house today.

36. William Diller Matthew and his daughter Margaret at the new home in Hastings-on-Hudson, 1912.

37. Dr. Matthew and his children at Hastings-on-Hudson, about 1915. *Left to right:* Margaret, William, and Elizabeth.

38. A family gathering at the Matthew home in Hastings-on-Hudson. Will and Kate are standing in front, second and third from the left. The ladies' dresses and hairstyles may give the reader some clue about when the picture was taken.

39. Kate and Will Matthew. Silhouettes by
Mrs. Lindsay Morris Sterling.

40. An informal portrait of Dr. Matthew in his office at
the American Museum of Natural History, about 1915. The
stray lock of hair and the almost inevitable cigar were trade-
marks of the man.

only for herself and others of the Lee family but also for friends and neighbors who commissioned her to be their seamstress. And she dearly loved to sew nice little things—handkerchiefs, scarves, spectacle cases, and the like—to be given as Christmas and birthday presents. Her skills obviously were considerable, so much so that during the winter of 1898 she was employed for several weeks at O'Hara's— "a swell dressmaker in N.Y."—where in March 1899 she returned to continue as a dressmaker.

Since she was a good seamstress and since she enjoyed the work, it was only natural that she decided to pursue it as a profession. On September 11, 1899, her diary reads

> I have about decided to go to Pratt's [Pratt Institute in Brooklyn] and take up dressmaking. If I take the whole course it will cost seventy five dollars. That does seem like a lot. Em has offered to loan me that amount which she calls "investment." I do hope I can take the whole course (a full year) as a certificate is given you at the end which Margaret Hayes (over at O'Hara's) told me was a great help. My idea is to go all this year (the course finishes in June) then rest next summer and next fall go to O'Hara's for maybe six months as her fitter and trimmer, then when I come away from there I can go to cousin Jo's friends and tell them I have graduated from Pratt, have been with O'Hara and am capable of doing their work.

The first part of her plan proceeded as she hoped it would. She attended Pratt Institute and received her certificate in May 1901. But instead of going to O'Hara's to work, as she had originally hoped to do, she stayed at Pratt as a member of the faculty.

On September 23, 1901, she "began work at Pratt. Fits like an old shoe." And there she stayed for several years, teaching other young ladies how to sew.

She was now quite a young lady herself, in her mid-twenties, and one of the young men in whom she became interested was her neighbor Alfred Diller. She saw quite a lot of Alf—the Dillers and the Lees frequently visited back and forth—and she confided to her diary that "Alf is *so nice*." That was one of many favorable remarks made about Alf.

The reader may recall that Alfred Diller and his cousin, Will Matthew, were particularly close friends, having been playmates in Brooklyn when they were little boys, and having renewed their ties when Will came to Columbia University.

On one of Will Matthew's visits to the Diller family in Brooklyn, on Sunday, April 27, 1902, Alf took his cousin over to the Lee home. Kate Lee's memo in her diary was "Alf and Bill Matthews [sic], P. M. to supper."

CHAPTER 9

Will and Kate

*T*he first meeting between Will Matthew and Kate Lee on that April evening in 1902, when Alf Diller brought his cousin over to the Lee home for supper, did not seem to strike sparks of scintillating interest between the two young people. It is to be presumed that a pleasant evening was enjoyed by hosts and guests, after which Will made his way back to his bachelor flat in Manhattan.

Perhaps other trips over to Brooklyn took place that spring, but whether Will saw Kate again is not known. That was the summer when Will went to Montana for his fourth western field trip, and it may be that the plans for the expedition were uppermost in his mind during May. However, it is quite obvious from Kate's 1902 diary (faithfully kept on a daily basis) that Will Matthew was *not* in her thoughts after their first meeting—nor was he for a long time. Indeed, her diary for the remaining months of 1902, and for all of 1903, does not mention Will (or Billy) Matthew. And there are no references to Kate Lee in the letters that he wrote to his parents.

Then in 1904 things began to happen. On May 27 Kate joined a group of young people—including various Dillers and "Billy and Hatch Matthew"—for a sail on Long Island Sound. It must have been a jolly party, for they "sang till eleven." This was followed on June

5th, a Sunday, by a visit from Billy, who "stayed till eleven. He's *mighty* nice." He appears ever more often in her journal for the month of June, which culminated with the entry of June 28, when she wrote that he was one of a large group of relatives and friends who gathered at a New York pier to see her off on a trip to Europe. She made the trip with three of her friends, including Will's cousin Helen Diller. They were gone until September 12.

In the meantime Will was off to Wyoming for the summer field campaign, but his thoughts were definitely also directed across thousands of miles to Europe.

> Fort Bridger, Wyo.
> August 16, 1904

My dear Kit:

> I wonder how this country would impress you. There aren't as many picture-galleries as in Europe and the cathedrals and ruined castles are mostly of Dame Nature's handiwork. Likewise there's no great abundance of trees, the nearest approach to them being sage brush and bunchgrass. Altogether, I'm afraid you wouldn't approve.

In this connection it may be interesting, especially to paleontologists, to repeat some comments about the field season of 1904 as set down in a letter written to his mother on November 12 of that year. His remarks throw some light on how the museum field expeditions operated during those early years of the century.

> The Professor [Osborn] was in Europe all summer and did not visit the camps, as he usually does. Brown brought back some interesting collections as he always does; he is the most successful of our collectors, and deserves his good fortune by hard work. Granger had charge of the Bridger expedition; I was out there only six weeks, and studying the stratigraphy during a good part of that time; but I was rather lucky during the time that I was collecting. I generally have a good deal to do with the planning and sending out of the expeditions; more this year than usual as Prof. Osborn left quite early in the spring.

Will (Billy) got back to New York in time to greet Kate (Kit) when she returned from Europe, and from that time on they were together as often as circumstances permitted. Such meetings were mainly on weekends, for they were both busy people—Will immersed in his museum work in Manhattan, and Kate teaching at Pratt Institute in Brooklyn.

One October Saturday Kate and her sister Emma were invited to spend the evening with the "Matthew boys," who had just moved into an apartment on West 145th Street, Manhattan. The boys, Will, George, and Harrison, had arranged a housewarming party, and of course Will wished to have Kate attend, which she did, properly chaperoned by her older sister.

Frequent meetings followed the housewarming party, including a dinner in Manhattan and a concert at Carnegie Hall. Then on Sunday, October 28, Will enjoyed dinner with the Lees, after which (having found a private place away from the numerous Lees) he made his marriage proposal to Kate. She accepted with delight.

Until those days in early 1904 when Will Matthew and Kate Lee became seriously interested in each other, Will had been very much a "man's man." His evening hours and his weekends were spent with his brothers or with other male friends, sometimes on bicycle excursions as noted. Frequently he went to Brooklyn to visit his Diller cousins, on which occasions he did keep company with Helen and Angela, as well as Alfred. He also visited Oscar Blackman who lived nearby. But except for his sisters Elsie and Bess, who at various times kept house for him, and the Diller sisters, he lived in a male world. Perhaps he was not a monastic type, but he certainly was not a young man-about-town, squiring an assortment of young ladies to evenings of entertainment.

As is so often the case with such young men, when he fell in love with Kate Lee it was a head-over-heels affair. From the year of their engagement to the end of his life he was completely devoted to Kate—his Kit or Kitty. She was equally devoted to her Billy—a name that she reserved for her exclusive use. (To the rest of the family and to friends he was invariably known as Will.)

Things were happening so fast in Will Matthew's life that some two weeks had passed before he got around to writing his mother about the engagement, although she had already heard of it.

Nov. 12, 1904

My dear Mother:

I had intended to write to you before about my engagement to Kitty Lee, but have hardly had a moment to write to anyone....

Also I knew that you were informed of the fact and so postponed writing until I had a spare evening.... I suppose I should be at work either on my report on the Bridger formation or on the guide that I am getting up for the fossil Carnivores Rodents and Small Mammals in our hall; but they'll have to wait till tomorrow.

I do wish you could meet Kit. She is the dearest sweetest little girl—a deal too good for me—and I have no right to such good fortune—but I must do my best towards deserving it. I hope we will be married before next summer but the date is not settled yet.

To balance Will Matthew's enthusiastic and perhaps not entirely objective description of Kitty to his mother, is an appraisal Oscar Blackman had written of Kate Lee—perhaps as a recommendation in connection with a position for which she was applying. He found her to be even-tempered, sincere, and extremely affectionate but also retiring and "rather too trustful," conscientious but sensitive to criticism, absent-minded, of a domestic nature, thoughtful, painstaking, but somewhat disorderly, and with strong likes and dislikes.

The question of what their shared future should be arose almost immediately after their engagement. The Carnegie Museum in Pittsburgh, a newly founded institution with the wealth of Andrew Carnegie behind it, was making a bid for Matthew's services. W. J. Holland was the director of the museum, and as a first step toward building up his paleontological staff he had, in 1899, hired Jacob Wortman (Matthew's immediate superior in New York) away from the American Museum. Presently he made an offer to Matthew.

The offer, evidently a most tempting one, came to Matthew in December, and he told Kate about it during his next visit to Brooklyn, on Sunday, December 11. One supposes that Will and Kate talked it over at length that afternoon. Would their future lie in the familiar surroundings of New York, or would they start their life together in Pittsburgh?

For Kate there was only one answer, and she expressed to herself in her diary: "I *pray* he won't accept."

The decision was his, and it didn't take him long to make up his mind. According to her diary three days after the Sunday in Brooklyn, Kate went over to Manhattan and "met Billy at 6:50. Had dinner at Thorpe's and he told me he had said 'no' to Mr. Holland!"

If it was a momentous personal decision concerning their future lives, it was also a momentous decision affecting the science of vertebrate paleontology in North America. Will and Kate were to spend the next two decades, and more, in the New York area, living in the home they would build in Hastings-on-Hudson, with Will at the American Museum. There they would enjoy frequent contact with various Matthew and Lee relatives as well as friends of long standing. For such family-oriented people Pittsburgh would have been a lonely place, at least initially.

One can only conjecture as to what might have been the result, in terms of paleontologic science, if Matthew had cast his lot with the Carnegie Museum instead of staying on at the American Museum. In either case his impact on North American vertebrate paleontology would have been profound, for Matthew was a giant figure in the field during the first three decades of the twentieth century. In Pittsburgh he would have been working with Wortman (at least for a few years), but with nascent collections of fossil mammals, whereas in New York he was a close colleague of Osborn (a relationship that had its ups and downs), and was able to work with collections that, even at the turn of the century, were of considerable size.

New Year's Day of 1905 fell on a Sunday and Billy (by this time Kate invariably used this familiar form of his name) spent the day at the Lee home in Brooklyn. And from that time on through the early summer he was at the Lee's almost daily: all day on Saturdays and Sundays, and evenings on weekdays. Occasionally Kate would meet him in Manhattan, after having taught all day at Pratt, and they would dine and perhaps go to the theater. But afterward he would, of course, accompany her back to Brooklyn, and then make the dreary trek back to West 145th Street in Manhattan. It was a strenuous schedule, to say the least.

In addition to socializing Will and Kate were making plans for the wedding and for a honeymoon trip. Will, with his characteristic attention to detail, developed an elaborate schedule for a journey that was to take them from New York to a European port, there through France to Switzerland where they would spend most of their time, then to London, from there to Saint John for a two-week vacation, and finally back to New York. The trip was to start on July 15 (ostensibly immediately after the wedding) and was to end on October 7.

Not only did Will plan each day, indicating places, distances, and elapsed times, but also he carefully estimated the expenses—day by day.

He wrote Kate "what a gorgeous time we're going to have on that trip. I do think I've worked out a bully route, and I hope it will like you as much as it does me. It may be reckless to blow in all our little margin on our trip, but it is something we'll look back to all our lives, and I think we can afford to take chances on it."

On an envelope containing the trip itinerary and his letter to Kate, there is a poignant little penciled note.

> March 20
> 1905
> Plans for proposed wedding trip. (Afterward given up and money
> lent to a member of the family. Wedding trip to St. John instead.)

How much disappointment is hidden in that note! Yet how much of the Matthew conscience, and duty to family, is revealed. The member of the family, whose needs were such that the idyllic vacation in Europe had to be abandoned, is not known.

The European trip hopes having vanished into thin air, Will informed his mother on April 21 that although an exact date for the wedding had not been fixed, it would be about July 15. He also said they expected to spend the rest of the summer in New Brunswick where, among other things, they would go on an extended canoeing-camping trip.

The wedding *was* held on July 15, 1905, in Brooklyn; the minister was Percy Olton, an Episcopalian clergyman who was a Matthew cousin by virtue of having married Theodora (known to the family as Dot), the daughter of Robert Matthew. (Robert was the brother of George Frederic, Will's father.) After the ceremony Will and Kate went to a hotel, and the next afternoon they took the Fall River boat to Boston, where they boarded a steamship for St. John. George Frederic Matthew met them at the pier and escorted them to the Matthew home where Kate met those members of the family who were presently in New Brunswick.

For several days Will and Kate shuttled back and forth between the George Matthew home in Saint John and a summer cottage known as Amkuk, located about fifteen miles up the Kennebecasis

River from the city. They were there on Monday morning, July 24, ready to embark on their much-anticipated canoe trip.

The canoe was docked at Fluwellings Wharf, near Amkuk, fully loaded with camping gear and provisions. The family, including several of Will's brothers, had gathered to give the newlyweds a hearty and affectionate farewell. At eleven o'clock Will appeared, properly attired for a camping trip in the Canadian woods. Then Kate appeared, dressed in an outfit she had made, an elegant ensemble (as might be expected from an instructor of sewing at Pratt Institute) consisting of bloomers (in those days the standard lower garment for athletic and out-of-doors young ladies), long stockings, tennis shoes, and a pretty shirtwaist set off by a necktie.

Kate stepped into the bow of the loaded canoe, which was being steadied by her father-in-law, who had a firm grip on the gunwale. As she lowered herself into the canoe George F. Matthew failed to allow for her weight; with a firm hand he held the bow high and, of course, the little vessel turned upside down, plunging Kate, tent, bedding, utensils, and food into the cold Kennebecasis tide. The water was shallow, there was no danger, but Kate was thoroughly drenched—except for her necktie.

Everything was rescued from the river and spread out on the grass to dry in the sun. Kate went back to the house for a complete change of clothes, and her nice outfit was also put out to dry. The departure was delayed for four hours, but finally they were off for three weeks of canoeing and camping.

After the canoe trip there followed weeks of sailing (all the Matthews were great sailors), camping, visiting, and feasting. The departure of the vacationing couple did not take place until September 23, when they took the boat to Boston and New York, where they began their life together.

Upon their return Will and Kate found a suitable apartment on West 147th Street in upper Manhattan, leaving the "Matthew boys," George and Harrison (and perhaps Charles), to continue their bachelor existence on West 145th Street. It was a busy life, with Will at the museum, with Kate setting up their new domicile, with numerous Matthews and Lees dropping in for visits, and with Will and Kate in turn visiting various Matthews and Lees.

From the beginning of their life together they were casting about for someplace beyond the city where they could live as they really wished. The place was to be Hastings-on-Hudson, just north of Yonkers and about fourteen miles north of the museum. A group of people, including some from Columbia, were getting together to purchase land in Hastings high above the river, and to subdivide the property among themselves. Matthew was one of the participants in this venture; so we find him as early as mid-November spending a long Saturday at Hastings, conferring with his future neighbors. In time the plan was successfully concluded with the result that Will and Kate became the owners of an acre of land (subsequently enlarged to an acre and a half), bounded on two sides by Edgar's Lane, an old, bending road where more than a century earlier there had been a sharp little Revolutionary War battle—on the site of the Matthew property.

Thus ended an eventful year for Will and Kate. Their separate lives had merged, and they were now looking forward to the years ahead.

CHAPTER 10

Their First Year

*W*ith the advent of 1906 Will and Kate were comfortably ensconced in their apartment on West 147th Street. They looked forward to a busy year, but only partly comprehended how busy it would turn out to be. Hastings-on-Hudson was on their minds. Moreover, they had to plan for Will's spending the summer in South Dakota to continue his part of the departmental field program—an activity he had interrupted the previous summer in order to marry Kate. Now the honeymoon was history, they were accustomed to their new domestic life together, and, as is so characteristic in the lives of paleontologists and their families, they were facing the prospect of a summer separation that is the frequent lot of the fossil hunter and the fossil hunter's spouse.

They had not, since the day of their marriage, been apart. Now they were to learn what it was like, and sooner than they expected—all because of the "Warren Mastodon."

In 1845, during an unusually dry summer, a complete, articulated skeleton of an American mastodon was discovered in the bed of a little dried-up lake near Newburgh, New York. The skeleton was purchased in 1846 by Dr. John Warren, professor of anatomy in the Harvard medical school. Dr. Warren had his own private museum in

Boston, and there, in 1849, he had the mastodon set up. There it remained until 1906, when it became available for sale, together with other fossils that Dr. Warren had accumulated in his museum. Professor Osborn learned of the availability of the Warren collection; he immediately got in touch with J. P. Morgan, who by telephone authorized the purchase of the Warren fossils. Matthew, by now very much Osborn's second-in-command, was delegated to go to Boston to supervise the packing of the Warren collection.

So on March 5, 1906, Will found himself in Boston, for the first time away from Kate, getting ready to dismantle the famous skeleton.

> I've had a busy day of it since early morning. Saw Prof. Dwight. He is a grandson of old Dr. Warren who made the collection, and is the fourth generation of professors of medicine or anatomy at Harvard, the fifth being represented by John Warren Jr., who is demonstrator of anatomy—a big fresh-colored, pleasant faced young chap, whom I liked very much off hand. They are all of them a nice outfit, these Warrens, with the kind of ideals that we set so much store by—of simple living, good health, broad interests, and accomplishing something in the world besides mere material things. Went over with [Prof. Dwight] and Dr. Warren to the old museum (which isn't open to the public, and hasn't any regular attendants) and have been nosing around and cataloguing all day. No heat in the building there and as you may suppose it's cold work cataloguing.
>
> The collection is disappointing, and I think more than ever that we paid far too much for it. However I'm not responsible for that and it doesn't come out of our Museum funds, but I know from some words that Dwight and Warren let drop that it's a good deal more than they expected to get for it. . . . The *Mastodon* is a splendid thing, very complete.   (letters to Kate, March 5 and 6, 1906)

His first view of the Warren fossils must have been less than cheerful. The collection on the whole was disappointing, dust covered everything, and the mastodon skeleton, although in itself an impressive fossil, was thoroughly blemished by a coat of thick, black varnish that had been liberally applied to all the bones; furthermore the tusks, intact when the skeleton was found, had disintegrated—fossil tusks are very fragile once exposed to air—and consisted of hundreds of fragments kept in boxes. A pair of papier-mâché tusks, much too long, had been inserted into the tusk sockets of the skull.

On the day after he had arrived in Boston Will received a letter from Kate. "It *is* the first letter you've written to me since we were married, isn't it? And to think how many things have happened since then! But I hope we won't have many more occasions when we have to write to each other, for even this short break is hard" (letter from Kate, March 6, 1906).

It was a vain hope, because as subsequent events were to prove (and as Will might well have foreseen if he had been looking toward the future with a realistic eye) there would be *many* occasions necessitating correspondence between them.

But for the present he was doing all in his power to make this first separation from Kate a short one. He worked day and night in Boston supervising the work, helping with the packing, and cataloguing the collection. He told Kate in his letter of March 6 to "be very sure dear heart that it won't be longer than can be helped by hard work."

On March 7 Will informed Kate that

> Dr. Warren has very kindly put me up at the Boston Athletic Association, but I don't know just how to take advantage of the privilege, as for the sake of good fellowship and getting the best work out of my assistants I have to eat with them and haven't any time for anything but work. But I appreciate the kind intent none the less.
>
> I wish I could see the end of this job more clearly. The packing will mean a fortnight's work, but I don't have to stay to see it through, but wish I could see the mastodon skeleton packed.
>
> (letter to Kate, March 7, 1906)

Things moved along much more rapidly than Will had expected, so that on March 11 he wrote to Kate to let her know he would be home the next day at about eight in the evening. This first separation, which lasted just a week, was at an end.

Kate had written to Will on March 6 that she had something to tell him when he got back home. If he guessed the secret (which seems likely) he gave no indication of it in his daily letters. Indeed, he said he was "on tenterhooks" to know what it was. Of course, what she had to tell him concerned the expected arrival of a baby in August, and she had the pleasure of revealing this news to him as soon as he came through the door of their apartment.

Life was getting complicated: a paleontological field trip beginning in June, packing their household goods and sending the lot to a warehouse in May, consulting with landscape and building architects before departure for the field, and preparing for a baby due in August. It was enough to make one's head spin!

Their first decision in the face of these complications, was for Kate to go to New Brunswick while Will was in the field, to await the arrival of the baby in the home of her in-laws. Then he would rush back from the western badlands, to be on hand for the birth. (One marvels at the toughness and resilience of young married couples.)

When Kate had departed for New Brunswick in late May, Will performed the task of packing the household goods—"five barrels, ten boxes, a great stack of parcels and all the furniture" (letter to Kate, May 31, 1906). Fortunately his brother George helped Will with this wearying and somewhat mournful work. "It is cruelly hard to tear up our little house, where we were so happy together," he wrote Kate on May 29.

But if they felt sad breaking up their first home, they experienced strong feelings of anticipation and excitement making plans for the new home in the north, high above the broad Hudson River. The Matthew acre in the subdivision of land acquired by the cooperative group of future residents was on the side of a hill. Will and Kate had ideas of placing the house on the upper part of their lot, overlooking the lower segment of their property where there was a little evanescent pond, enclosed within masonry walls and fed by an intermittent rivulet. This pond left over from Revolutionary times, caught their fancy—could it be retained and maintained as a desirable feature of their land? They decided to seek professional advice. Will informed Kate that "Mr. Thayer [the husband of Kate's sister Emma] called me up this morning and commended me to Mr. Downing Vaux, in the matter of getting a plan for the improvements at Hastings. So I made an appointment to see Mr. Vaux tomorrow" (letter, June 4, 1906).

Downing Vaux was the son of Calvert Vaux, who with, Andrew Jackson Downing and Frederick Law Olmstead, dominated American landscape architecture during the latter half of the nineteenth century. Downing Vaux was a lesser figure than his famous father (who had played a crucial role in the planning and development of Central Park in Manhattan and Prospect Park in Brooklyn), yet he must have had good advice to offer on the relationships of structures and land, based upon parental precept and example.

On June 6 Will went over the ground at Hastings with Mr. Vaux, who promptly scotched the idea of retaining the pond, pointing out that the rivulet feeding the pond was insufficient to keep it "clean and wholesome" and that the pond would not be of much use. The pond *did* seem romantic in prospect, but the hard facts had to be faced, so Will advised Kate "it is somewhat of a disappointment that our pond proves impracticable, but I feel that we would be very foolish after having called in expert advice, not to go by it, and should not hesitate to accept his opinion on a practical point where he has had a great deal of experience and we none. He seems to be a mighty capable man. As for the little house he entirely approves the position we picked for it" (letter, June 6, 1906).

The consultation, which involved the better part of a day, what with travel between New York and Hastings and several hours at the site, cost the Matthews all of twenty-five dollars.

During those days of packing furniture and surveying the land in Hastings, Will Matthew was, of course, still busy with his duties at the museum.

> Our artist is making a restoration of the skeleton of *Tyrannosaurus,* the gigantic carnivorous dinosaur, based on my preliminary restoration which had a skeleton of a man to the same scale in the corner. The big dinosaur is standing on his hind legs and holding out his little clawed hand for all the world as though he wanted to shake hands with Mr. man, and with his big row of teeth and all, there was a suggestion of Roosevelt about him. And the man is holding his hands partly backward, looking as though he didn't particularly want to shake. So I scribbled underneath the drawing.
>
> Teddysaurus—Delighted, I'm sure.
>
> The Common People—Hum, do I want to shake hands or not?
> The artist saw this and was terribly tickled.
>
> (letter to Kate, June 1, 1906)

By June 12, Will Matthew was in Omaha on his way to the western badlands. Loose ends at the museum had been tied up for the summer's duration, the household goods were now reposing in a warehouse, and, most important, Kate was comfortably ensconced in New Brunswick, away from the heat and grime of a Manhattan summer. Will was now free to give full-time attention to the fossils of South Dakota, almost—every evening he usually wrote a letter to Kate.

His first stop was Rapid City, South Dakota, where he was to join Albert (Bill) Thomson, his partner for the summer field campaign. Thomson was a native of the region, having grown up on a ranch near Rapid City. But as a young lad in the nineties he had become attached to an American Museum field party, which led to his permanent employment in the paleontological laboratory of the museum. At an early stage in their careers Matthew and Thomson became closely associated as collectors, preparators, and students of fossil mammals. The partnership was to last for as long as Matthew remained at the American Museum.

The two friends met in Rapid City on June 14, and there they took the initial steps for their summer of fieldwork, including hiring a cook, assembling their field equipment, and providing for transportation. The latter consisted of two teams of horses and two wagons— a light team with a light wagon for everyday use, and a heavy team and wagon for the strenuous hauling.

To reproduce all of Will's letters to Kate, or even a large sampling, might make tedious reading, but selected letters give an interesting picture of the work during that summer. Also, these letters' descriptions of camping under primitive conditions and the logistics of horse and wagon transport offer a striking contrast to the comfortable camps, modern equipment, and motorized transport of fieldwork today.

> Rapid City, S. Dak.
> Thursday, June 14, 1906

Well, I got in last night to Rapid, and am installed with Thomson and his sister in his father's shack in town. The old man lives on his ranch about 35 miles from here, but keeps this place in town as well; a great convenience for us as we don't have to stay at a dirty little country hotel. Thomson hasn't yet succeeded in getting his outfit together, but hopes to have it inside of a couple of days, and then we will start south and east, around the southern end of the Big Badlands and then up on the tableland to the east and south of the White River. Our next postoffice address will be Manderson, S. Dak.

Thomson's Ranch
June 18, 1906

We have got as far as the mouth of Spring Creek, near the Big Cheyenne River and have been getting the remainder of our outfit into shape.

Saturday we started out from Rapid, and got about 17 miles, when a big storm came up. Started out with hail—some of the hailstones over an inch in diameter, so we tied the horses to the wagon to keep them from running away; then a short let up, and we started on again but had not got far when it commenced worse than ever. So we made for a ranch on the road and just got our horses in when it came down in bucketsful. Stayed there over night, laid our beds in the loft above the stable. Storm was pretty fierce— more thunder and lightning than I've seen for some years. One flash struck near us and put a telephone wire out of business— splintered up posts and burnt up the wire. On the railroad it blew a car with about a dozen people in it off the track into the ditch, and damaged some of them more or less.

Sunday morning we started on again and got to [Albert] Thomson's father's ranch about dinner-time. Have been busy since then in scraping, oiling and painting our little wagon, resorting and loading the heavy one and picking up the various odds and ends that we will need on the way. For cook we have an old chap, more or less crippled with rheumatism, but a faithful hard-working old fellow, who can do what we need to have done—cooking, looking after the horses, keeping around camp and so on.

On the whole I think we have a very good outfit. The light wagon, which Gidley and I bought second-hand some three years ago, was very disreputable looking, but the running gear is sound and good, so we took scrapers and scraped it all clean, and then oiled it all over, body, wheels and gear, with linseed oil, and then painted it with carriage paint—Brunswick-green, that is to say, very close to black. It goes by the name of the undertaker's wagon, but it will serve our purposes just as if it were bright red and blue touched up with green and yellow. I don't know how much our small team can do in the way of work, but they look well and are fat as butter. They have been running on the range all summer and may be given to circussing round more or less. The heavy team, which will have the big wagon to pull, are first-class horses, and well worth what we paid for them as prices of horses go now. We will use the light team to pull our light wagon and also as saddle-horses, and will probably also take my old saddle-horse—Kid—if he is any use. He has been cut up some by getting into a barbed-wire, and may not be fit to use—if he is, we'll take him along.

We have a tolerable supply of guns and ammunition and may come across a few rabbits to help out our cuisine. There isn't any large game much in this country, and if there were, we haven't time for hunting. So all we need is a little target rifle and revolver. Thomson has also a larger rifle—30.30—but I don't think he'll use it. . . .

Thomson's father is a typical old ranchman, a fine old chap who has lived in this country ever since he came over from Denmark. His mother is a nice little old lady, and there are two sisters, one older, one considerably younger. The latter is something like Ethel in looks and manner, and just out of a finishing school at Spearfish, S.D. The ranch is like most of them in this country—buildings and barns and corrals scattered around higgledy-piggledy over two or three acres of ground, with a very pretty little stream which carries water all summer and is used a little for irrigating. No garden and no pretense of neatness or order. It impresses you so much that with a little trouble and thought it could be made a very nice looking place and, as it is, it is excessively ugly, although comfortable enough to live in.

Various neighbors drop in from time to time, and sit around and telephone in the house—a party line that serves over a dozen people in the neighborhood—keeps going pretty steadily from five a.m. to ten p.m. The old chap is a hospitable and generous old fellow without any frills, but very likable. Albert is evidently a favorite among the young men around, and they drop in and sit around and gossip about the neighbors *ad lib.*

Wounded Knee Cr'k [Pine Ridge Res.]
June 21, 1906

This is the first chance I have had to write a line since we left old man Thomson's ranch. We have been on the road all day, and by the time supper was over and the horses attended to it was too dark to write. But tonight we are a little earlier so here goes:

We left the ranch on Tuesday and had travelled about two miles when another storm of hail and heavy rain drenched us thoroughly. We pulled through the rain 'till we got across the Cheyenne River, and then could travel no further, as our road lay up Indian Creek, and the storm had preceded us up that creek and filled it bank full, so that there was no fording it. So we stayed over night in a little shack that Albert Thomson has built on a quarter-section of the land that he owns on the bank of the Cheyenne. We set up our stove and changed our clothes partly, and dried ourselves out.

Next morning we started early, getting up about half-past four and on the road at seven o'clock. We started up Indian Creek, crossing and re-crossing it on a very bad trail, as the creek is continually changing its course, and a good ford one day may be a steep over-hanging cut bank after the next heavy rain. The rain of the day before had spoiled a lot of the crossings and we had to hunt new ones. There was very little water running in the creek, but it had been three feet deep the day before. It drains a lot of White River badlands and when full it is a thick mud, almost impossible to settle. This mud had been running into the Cheyenne the night before, so you can imagine what a nice combination we had to drink, of Cheyenne R. alkali and Indian Creek mud. We didn't need any milk in our coffee; it looked like rich café-au-lait, without anything in it. Well, we pulled up to the head of Indian Creek, and crossed the divide at Sheep Mountain and came down Cottonwood Draw to White River. This was in the Indian Reservation, and as soon as we came down to the river we found lots of Indian ranches scattered along it. We stopped back of Yellow Bird's ranch, where there is a fine, cold, clear spring. It had been blowing hard from the north all day, and was quite cold at night. The divide between the Cheyenne and White Rivers is a mass of badlands, and points and buttes come nearly down to the rivers on each side. These badlands are of the White River formation, and are not the one in which we want to collect. Our work lies in the Loup Fork formation which overlies the White River beds. We had travelled about thirty miles or more on a pretty bad road, and by the time we got in at night it was half-past six. Then we had to unpack, get supper, water and feed the horses and picket them out for the night, and then went to bed.

This morning we were up a little before sunrise, and on the road at 6:45. We crossed the White River and pulled up Wounded Knee Creek. We came up to Manderson and got our mail, but had not time to stop, and came to this place, about five miles above Manderson, the first place that we could pitch camp. The creek flows through a fine rich bottom, but it is fenced in all the way, the road mostly running between two fences, and there was no chance consequently to get at the creek for water and at the same time have grass enough for the horses. On the hills on each side of the creek are small patches of badlands or exposures of rock, White River formation below, Loup Fork formation on the upper part of the creek. We reached here about two o'clock, had lunch and started out to hunt fossils. It isn't a very rich section, but there are some fossils, and Thomson and I each found a skull of *Merycochoerus,* a large ruminant which is rather rare. In our collections are several skulls but none from this region. Came back about seven, had dinner and began this letter. It is getting so dark now that I can hardly see to finish it.

This letter describes in vivid terms the slow and rugged nature of badland travel back in those days before jeeps and field trucks, something that perhaps is not always appreciated by today's generation of paleontologists. Moreover, one is apt to forget (if one was ever aware of the fact) how much time and attention the field naturalist had to devote to finding a campsite where there was grass and water for his horses, and to otherwise caring for the animals.

By the time Will had written the following letter, he, Bill Thomson, and their cook, Tim, were getting settled in the daily routine of fieldwork. For the rest of the summer their days were spent searching for and collecting fossils in the vicinity of Wounded Knee and Porcupine, about ten miles apart some sixty miles southeast of Rapid City. Each day was more or less like the next, but they were not monotonous. Will provided Kate with a description of a typical day.

Porcupine Creek
July 8, 1906

This being Sunday I'm going to write you a good long letter, as we are laying off for the day. It is nearly noon, and I'll have time only to commence it before dinner but after dinner I can continue it at length. There is an old Indian in the tent now, an old, old chap, pretty deaf and doesn't speak any English, but a very decent fellow, keeps himself neat and clean, which is more than most of them do, and has quite taken to us. He drops around about mealtimes, as they often do, gets a 'hand-out,' and says he knows lots of bones but we haven't been able yet to make out just whereabouts he means. It may be in the big badlands to the west of here, in the White River formation, and not of any special interest to us. We'll try to find out.

Well I don't think I've told you much about our camp life. We have a 10 × 12 tent, and a sheetiron camp stove, and at present a pretty good supply of firewood, though it isn't always easy to get any as the Indians clean out pretty much all the dry wood they can lay hands on. We have a heavy wagon for our camp outfit and a light one which we use every day to go to and from our work. Also we carry home our specimens and firewood in them. Each morning Tim, the cook, gets up at dawn and gets breakfast ready by the time we are washed and dressed. Then we have a breakfast—a pretty solid one, too, no cup of coffee and shredder for us out here. Then we light our pipes, catalogue and finish pasting our specimens if we have any to do up, feed and water and harness up the horses and get off some time between seven and eight. Get down to our work and unharness and picket out the horses and start hunting specimens, or prospecting those we have already found. You find a specimen, the back of a skull or a few fragments of bones sticking out of the rock, and chisel and pick and scratch around it to see what it is, then cut away the rock from above so that you can see whether there is anything left in the rock. Having found out the lay and limits of your specimen you channel around it and soak it with a thin solution of shellac in wood-alcohol, to harden it, then cover it with burlap strips dipped in flour paste to keep it from breaking apart.[1] Then you cut underneath until you can break off and lift the block and turn it over to paste the under side. It takes some time to dry the paste and while it is drying you go and hunt more specimens. Of course in most cases the fragments sticking out of the rock are only fragments or there isn't enough left to be worth bringing in, and you have all your work for nothing. About six o'clock we come back to the wagon, harness up and start for camp. If we are near enough we come back at noon also, for lunch. We always take a couple of canteens of water and package of lunch with us. These, with our shellac bottle, chisels, awls, wrapping paper and cotton, twine; brushes for removing dust and fragments from the surface of the bone, and can of flour paste and strips of burlap we carry in our "war-bag" on our backs, roped over

[1.] Today's paleontologists do not use flour paste for bandaging specimens; plaster of paris is the preferred agent. Various modern chemicals rather than shellac are used for hardening fossil specimens.

the shoulders. We get back to camp between seven and eight o'clock at night, and get supper, smoke a pipe (and write a letter if there is any light left) and then turn in. How's that for a good day's work? You see one has to be thoroughly interested in the work to keep it up on such long hours, and if no specimens turn up for a long time it gets pretty tiresome. But when you do find a specimen—some new species or missing link, or the skull or skeleton of some animal of which only fragments have been found before—it makes up for all the hunting and you feel that you've made a solid addition, permanent even if small, to scientific knowledge, and it's worth while after all.

Thus far we've done pretty well. About fifty numbered specimens up to date, and I should regard 150 specimens as a successful season's haul. If we go on at present rate we ought to make that number or more by the first of October. . . .

Tim is a regular camp cook—makes pretty good biscuits for camp, and to him the one necessity for every meal, breakfast, dinner and supper is biscuits and bacon. Other things are extras, but these there must always be. As the saying is, I've eaten that much bacon I'm ashamed to look a pig in the face. And his coffee is made by putting on the pot—old coffee & grounds, adding some more coffee and water and boiling it to death. The first thing for you to fix up for me will be a cup of real Kitty coffee, so I can see what it tastes like. After the way you've tuckered me up and coddled me last winter I'm getting to be fussy about food. I tried to show Tim how to make macaroni a l'Italienne, (he had never seen macaroni before) and made a dish for supper one day. He makes it now, but you'd never imagine what it was intended for, or that it was the same dish. It's a truly wonder how he can spoil good food. Never mind, anything goes when you're camping, and I'm not going to hurt the old fellow's feelings by nagging him about his cookery. He sure does the best he can and is civil and willing to do all in his power.

We have four or five boxes of specimens together now, and are using our wagon for storing a lot of the blocks as there isn't room in the tent for them. There is a sawmill on the creek a few miles below here and we can get some lumber there and make boxes before we leave. We may have ten or a dozen boxes to send in from here before we get through. I hope so.

Although Will Matthew mentioned that it might be possible to accumulate 150 good specimens by the first of October, he certainly did not expect to be in camp to celebrate the attainment of that goal. It had been arranged, even before he and Bill Thomson had gone into the field, that he would leave in late July, so that he could be present for the birth of the baby in August. He anticipated that event with pleasure and in mid-July was writing to Kate letter after letter expressing his impatience to be off and away. In spite of successes in the field (Matthew and Thomson had discovered some superb fossil mammals) he felt almost trapped during those last two weeks of work.

Finally on July 27 he made his departure, leaving Bill in charge of the work for the rest of the summer. Thomson was not to be alone; William King Gregory had joined the expedition in order to get some field experience. As Matthew left he felt certain qualms about Gregory. Would this young man, a scholar in the classic sense of the word, be able to cope with camp life? Could Gregory, a habitué of library and laboratory, deal with the rigors of the fierce, dry heat of the badlands in August? Gregory did, but he did not go on to become a field paleontologist.

Will arrived in New York by rail, and immediately took the Boston train, and then the boat for Saint John, arriving in the evening of the last day of July. It was a joyful reunion. For Will it was the end of a long separation spent in hard physical work under the blazing western sun; for Kate it was the end of the same long separation spent in waiting, the tedium of which was alleviated by her in-laws, and in experiencing the last months of her first pregnancy. Now, for the two of them, there would be late summer days together, visits by the hour, walks, little tasks, much reading and the anticipation of the arrival of the baby.

Will thought it would be nice if the baby were to be named William Pomeroy. *She* arrived on August 27, so the baby was named Elizabeth Lee. For about a week Will admired this newest Matthew, and then, on September 4, he boarded the *Calvin Austin* bound for Boston. He had been away from the museum for almost three months, and many matters awaited his attention in New York.

It is ironic that Will Matthew, who had spent two months in the badlands enduring heat, aridity, thunderstorms, and physical exertion, the Will Matthew who had left South Dakota worrying about the health of Will Gregory out in that wild country, should return to New York and an illness that brought him close to death. As soon as he arrived in New York he complained of having caught a cold. Thereafter his physical condition deteriorated rapidly, and for three weeks he lay in bed, extremely ill.

He was debilitated by a high fever, which he endured during a typical New York September heat wave, as usual accompanied by high humidity. Furthermore, the treatment he received from doctors and nurses would appear to have been somewhat on the Spartan side. Will wrote Kate he had "cold baths every hour, mustard plasters and several kinds of nasty medicine." However primitive such treatments might seem today, it seemed to do the trick; he was relieved of pain and his temperature subsided.

For a while, however, he was so weak that his cousin Helen Diller had to write his letters for him. All this alarmed Kate, and she cut short her sojourn in Saint John, packed her bags, and, with the help of Eloise Tyler, the fiancée of her brother-in-law Charlie, boarded the *Calvin Austin.*

By the time Kate and the baby arrived in New York Will had passed the crisis of his illness, but he was still very weak. On September 24, he was moved from Manhattan, where he had suffered for those intermidable three weeks, to Kate's family home in Brooklyn, arriving there "in George's arms." It was his brother George who had watched over him during the anxious days in Manhattan.

The doctors, and there were several, could not agree as to the nature of Will Matthew's illness. On September 27 they drew three pints of liquid from his lungs. Additional fluid was drawn on September 29. Perhaps he had suffered from pneumonia followed by pleurisy.

Will recuperated in Brooklyn, and by October 15 had recovered sufficiently to make his first visit to the museum. The next day Will and Kate decided to rent an apartment—as she noted in her diary— "at the corner of Bradhurst and 149th Street" in Manhattan. (Bradhurst Avenue is a short street bordering the east side of Jackie Robinson Park—formerly Colonial Park—which is at the foot of the bluffs that separate Harlem and Washington Heights from the Harlem River lowlands.) Five days later they moved into this, the second of their homes, where they waited in an empty apartment from nine in the morning until three-thirty in the afternoon when the van finally arrived with their furniture. Here, with the infant Elizabeth and a servant girl named Molly, they were to live for six months.

CHAPTER II

Hastings-on-Hudson

*W*ith the birth of Elizabeth in August and Will's illness in September, little thought had been given to Hastings. But once the family was settled in the apartment on Bradhurst Avenue, with Molly on hand to help at home, Will and Kate were able to resume planning for the home by the Hudson River.

Will was back at the museum, sufficiently recovered from his illness to resume his curatorial duties and research projects in his accustomed dedicated manner. Indeed, he spent many of his evenings at home "writing hard" as Kate wrote in her diary; writing, it would seem, some of the more philosophical parts of his great monograph—"The Carnivora and Insectivora of the Bridger Basin"—which was later published as a large format Memoir by the museum in 1909. (When it appeared Kate struggled through a few pages, as she felt a dutiful spouse should, but soon gave up in despair and informed Will "I'll cook your meals and I'll wash your clothes, but I will *not* read 'Carnivora and Insectivora of the Bridger Basin.'")

Kate was occupied, of course, with the baby and with domestic chores yet spent many evenings talking with Will about the house of their dreams. The precise site of their new home had been determined by Will and Downing Vaux, the landscape architect. On February 18, 1907, Will and Kate went up to Hastings where they found a house for rent at a location known as Tower Ridge. They decided to take it as a place to live while the house was being built on their lot. Now things began to move along. On April 15 they moved from their Manhattan apartment to the Tower Ridge house, which was to be their temporary home for a couple of years.

Will and Kate worked to make their rented place as comfortable and homelike as possible, even though they knew full well that their tenancy would be relatively short. Kate put in a lot of effort cleaning and painting and hanging curtains, and Will tended vegetable and flower gardens. And when they had a few hours to spare, they would climb the hill to inspect their lot.

Thoughts of a house of their own occupied their autumn and early winter evenings at Tower Ridge, and they had been consulting William Sanger, an architect who lived in Hastings. On January 18, 1908, they "suddenly decided to get Mr. Sanger to go ahead with the house plans."

William Sanger was the husband of Margaret Sanger, who, during the early years of the century was very active—and achieving a national reputation—in what today is known as family planning. In those days it was called birth control. She was a vigorous crusader for the dissemination of knowledge about contraception, and, because of her well-publicized campaigns, she was frequently in and out of Manhattan jails and repeatedly appearing before Manhattan judges. As Mr. Sanger worked on the planning of the Matthew house he and Margaret became increasingly close to the Matthew family. The Sanger house was just up the street from the site for the Matthew house, so they saw a good deal of each other.

Then, on February 20, Kate wrote in her diary that there was a "dreadful fire at the Sanger's last night. Swept the place clean. I went up to see if I could do anything—so sorry for them."

And Kate and Will stood by the Sangers in a big way—on February 29 they took them into their Tower Ridge house for the month of March. The Matthews rummaged in their closets for clothing, so that the Sangers would not be completely without resources in those first days after the fire. And on April 21 Kate and Will withdrew money from their savings "to help the Sangers buy off the vultures"—whatever that cryptic remark in Kate's diary may mean. So friendships were cemented to the degree that Margaret Sanger and Kate Matthew were close all during their lifetimes.

William Sanger planned and supervised the building of a commodious house on the Matthew lot. It was a two-story structure with a separate apartment on a third (attic) floor consisting of a living room, another small room, a bedroom, and a bath—all intended as a maid's quarters.

The house had a stucco exterior finish, and there were two large porches at the ground floor level, and the roofs of these porches formed second-floor decks. There was also a little porch off the attic floor living room. And there was a large basement. A notable feature indoors was the living room, with on one side a fireplace flanked by pale burnt-orange tiles that extended up to a large, shallow wooden arch spanning almost the entire length of the room; the arch terminated at each end on wooden pediments supported by carved wooden dwarfs or elves. The wall opposite the fireplace had, beneath a matching arch with carved figures, a large window with a broad window seat.

It was basically a nine-room house. On the first floor were, in addition to the living room, a large dining room a kitchen, and two pantries. On the second floor were three bedrooms a bath, and a study. The third floor, or attic, originally housed a living-bedroom—intended for a maid—and two storage rooms, which eventually were converted into a bedroom and a kitchen. Thus the third floor became a self-sufficient apartment that in later years accomodated Kate's sister, Mame, and their widowed mother. There was a little third-floor balcony, from which the Matthew family viewed the total eclipse of 1924, and from which they set off fireworks on the glorious Fourth—this in spite of Will's Canadian heritage. The basement housed a big coal-fired furnace, a laundry room, a work room, and a large storage room lined with shelves—these stocked with jars of preserved vegetables and fruit from Will's garden.

The house was completed and occupied in ample time for the arrival of Margaret, who was born in Brooklyn on April 18, 1911. William Pomeroy Matthew, the name chosen for the first Matthew child, only to be set aside upon the arrival of Elizabeth, and then of Margaret, was finally bestowed on a son born September 18, 1914.

The Matthews lived happily in this gracious house for seventeen years, a period that saw many family gatherings, especially on Sundays, of Matthews, Lees, Dillers, and other relatives and friends.

Will Matthew's long-term and lasting contribution to the amenities (defined in the Oxford dictionary as "the pleasurable features of an estate") of the Hastings place was his work on the grounds. He loved gardening and landscaping, so through the years he built terraces (with shovel, hoe, and wheelbarrow), planted flowers and trees and vines, and down on the lower part of the lot cultivated an extensive vegetable garden. Indeed, so efficient was his work in his vegetable plot that it produced all the vegetables the family needed during the summer months, with enough surplus so that Kate was able to can quantities of beans, peas, tomatoes, and the like for winter consumption. Carrots and beets were stored for winter use in boxes of sand, and the fruits of the trees and vines were canned or made into jams and jellies. For a garden on suburban estate, it provided a considerable measure of self-sufficiency.

This was Will's form of recreation; he never felt the need to belong to a country club. It was his habit in the warmer months to arise with the sun and work for an hour in his garden, then come in for a shower and breakfast, after which he would walk down the hill to the station where he boarded the commuter train for the city. As if his stint in the garden wasn't enough, he would, in good weather, disembark from the train at 125th Street and Park Avenue and from there walk downtown and across Central Park to the museum—at Central Park West and Seventy-Ninth Street. This excursion involved a hike of about three miles on hard city sidewalks, an exercise fatiguing even to contemplate. But Matthew, the erstwhile "sickly" child, had developed a life-long devotion to strenuous exercise.

In this respect he was somewhat akin to his contemporary, that life-long friend and patron of the American Museum of Natural History, Teddy Roosevelt. President Roosevelt, was, in fact, quite aware of some of Matthew's scientific work, as indicated by a letter he wrote to Matthew in 1915 in which the ex-president expressed satisfaction in seeing "Climate and Evolution," just off the press.

May 3rd 1915

My dear Mr. Matthew:-

I am immensely interested in "Evolution and Climate." It seems to me one of the best pieces of work that has recently been done. I am not yet satisfied that all of evolution can be accounted for by external stimuli, however. The first chance I get I want to have the opportunity of seeing you.

Sincerely yours,
Theodore Roosevelt

Even though Will, like Teddy, was a zealous advocate of an active strenuous life, he also spent quiet hours at home and in the field with books and other reading material.

His study in the Hastings house was not a place dedicated to scientific reading and writing; these activities, he believed, were more properly pursued in his museum office. Moreover, his study was often occupied by some visiting relative. So he read at home in a comfortable living-room chair, purely for relaxation, and commonly with a cigar to keep him company. Of course he read the daily papers in those preradio, pretelevision days, and the *Illustrated London News* always lay on a table by his chair. Interestingly, he found escape from the world around him in pulp fiction; he was an avid reader of *Argosy All Story Weekly,* and he enjoyed western fiction. Matthew liked strong cigars and strong coffee; indeed, his love of these two stimulants was legendary. In his office and in the Hastings house his devotion to cigars was almost Churchillian. Yet he gave up smoking for one month each year—in part to prove to himself that he was not a slave to nicotine. As for coffee, his consumption of this beverage was anything but casual. It was his custom periodically to go down to the docks in New York, where imported coffee was unloaded, and there he would purchase a large burlap bag of green coffee beans. He would lug this home by subway and commuter train and then roast some of the beans in the oven of the kitchen stove. He would process the beans in small batches in order to always have freshly roasted coffee beans at hand, and each morning he would grind his daily allotment in a small hand-operated mill. Only coffee obtained and prepared by these methods was suitable for his palate.

Another Matthew custom was his enjoyment of a glass of good beer with the evening meal. But when Prohibition became the law of the land after the First World War, Will Matthew gave up his evening glass of beer. He felt that he must obey the law, even though the law was, in his opinion, wrong and silly. This deprivation he endured without complaint for the remainder of his life.

During his frequent summers in the field, Will had a gardener who kept the place in order. Many of those summers the rest of the family went to New Brunswick for their vacation, there to be joined by Will at the end of his season of fossil-collecting in the West.

The Hastings house was visited by many paleontologists from various institutions in North America and Europe, and was a place for relaxed encounters and unhurried conversations. Needless to say, these visitors appreciated the pleasant hours spent up the Hudson River, away from the bustle of the city.

There were good times and bad for the Matthew family in their home in Hastings. Fortunately the good times were in the ascendancy and constitute most of the family remembrances. But one of their bad times was the terrible winter of 1917–18, a winter when trolley cars in various parts of Manhattan froze in their tracks and often remained immobilized for weeks, a winter of a great national coal shortage, a winter so cold that Will Matthew could pull his daughters, Elizabeth and Margaret, on a sled across the ice-bound Hudson to New Jersey and back. It was the winter that a great influenza epidemic swept across much of the world. The entire family was struck down, and all of them would have been in desperate straits had it not been for the life-saving efforts of Dr. Tyler, the father-in-law of Will's younger brother Charles. Dr. Tyler, a retired physician, came to Hastings and lived with the family, attended their sick beds, administered medicines, nursed the invalids, cooked their meals and fed them, tended the furnace, and generally kept the house in order. As the patients recovered, one by one, he placed them in front of the fireplace, and fed them on tea trays. And so, thanks to him, they weathered that crisis.

Another bad time came the following summer with the death of their new baby daughter, Christina, born the previous year. She was a belated victim of the influenza epidemic of the past winter.

On the other hand, serving to alleviate the bad memories of the previous winter, was the golden wedding anniversary of George Frederic and Katherine Matthew, celebrated on April 1, 1918. The elder Matthews came down from Saint John for this event. The various Matthews and Dillers, so far as they were able, assembled at the house in Hastings, and there the younger generation—Elizabeth, Margaret, and their cousins, together with their aunts and uncles, performed an operetta, "Rip Van Winkle," composed by Ned Manning, the husband of Will's musical sister, Elizabeth.

Will, on behalf of his siblings, wrote "The Golden Wedding," a series of verses that were read to his parents.

(The Children Speak)
>Today we have come together, gathered from near and far
>To celebrate your wedding date and the fifty years that
>>have passed.
>A golden record of honored lives, stainless and sweet and
>>clean,
>An active share in the world's best work, and a host of
>>friendships fast.
>To us, your loving children, you have given the best of
>>your life:
>It is ours to maintain your standard, and the honor of
>>our race.
>What we have done with the heritage you handed down to us—
>How we have used our talents, each in its proper place.

(Elsie speaks) [The unmarried daughter]
>Mine is the dearest privilege,
>And nearest of all to your heart—
>To keep the home fires burning
>I have taken the daughter's part.

(Will)
>It is mine to continue my father's work
>And search in the leaves of stone
>To decipher the records of worlds long past
>And to make their story known.

(Bessie) [musician, married to Ned Manning, composer]
>Your love of music, oh mother mine,
>Is the part I chose for my own—
>To play, to teach and to help my man
>To compose the songs you have known.

(George) [teacher and schoolmaster]
>A noble talent you gave me
>To employ as best I could,
>I teach my boys as you taught me
>To be wise and strong and good.

(Harrison) [business executive]
>To do my share in the business world
>Was the part assigned to me;
>But now I have answered a higher call—

To defend our liberty.
I have taken the place of my brother who died
In the battle across the main
And I've vowed to do my uttermost
That he shall not have died in vain.
(Robin) [killed in battle at Ypres, 1916]
I am with you in spirit here today,
Though you cannot see my face.
I have cast no shame on your honored name
But have done as became my race.
(Charlie) [in agricultural business]
I have taken my part in the business mart
And the talent you gave to me
Aids the farmer to make his wide-stretched fields
Green with fertility.
(Jack) [also in agricultural business]
I too am aiding the same good work
And giving the best in me
To help in feeding a starving world
Here and across the sea.
(The grandchildren speak)
Our look is toward the future, we still have the learner's
part
To fit ourselves for the coming years, in body and brain
and heart
That when we take up life's burdens, we may all of us do
our share
And our lives like yours, be an open book, with its pages
clean and fair.

Three years after this happy family celebration the parents of Will
Matthew came to Hastings to live, and there they died, in 1923. His
father passed away on April 14; his mother about a month later. The
next year Kate Matthew's mother, Jane Lee, and sister, Mame Lee,
came to Hastings to live, and occupied the third floor apartment.
Grandmother Lee died during the winter of 1926–27.

The Hastings house was truly a haven for the Matthew tribe, and
they treasured it as the center for their family activities. It had a
warmth imparted to it by a loving, close–knit family, and there they
gathered in evenings and on weekends, to escape for a while the cares

of the outside world. This was especially true for Will Matthew; when he came home he left museum concerns behind him. The intramural problems of the institution, which were becoming increasingly diffi-cult for him during the early twenties, were seldom mentioned or discussed, even with Kate, and that is how he wanted it. He wished his home to be a haven.

CHAPTER 12

Matthew and Osborn

*W*hen Will Matthew began his paleontological career at the American Museum of Natural History in 1895, he entered into a long personal relationship with Henry Fairfield Osborn. At the beginning this was a relationship of student and teacher—a continuation of those final years of Matthew's graduate work at Columbia, the years when he changed his contemplated profession from mining geologist to vertebrate paleontology. The teacher recognized the potential of this bright, young Canadian; the student, surrounded by fossil bones and intrigued by what they had to tell him about life in the past, appreciated the learned guidance he received from his urbane, worldly professor. Thus, in those final years of the nineteenth century, Osborn was the man in charge, and Matthew was his efficient lieutenant.

However, their relationship, especially in the early years, involved more than the academic kinship of teacher and student. There was a difference in age, Osborn being fourteen years older. More important, there was a sharp difference in their personalities, whether inborn or the result of the very different environments they grew up in.

Osborn was a privileged child of wealth, accustomed to having his needs seen to by servants and retainers of the family. He grew up in the family home on the Hudson River, a castle really, high on a hill opposite West Point. He attended exclusive schools and then went to Princeton. His father, a railway tycoon, fully expected his son to enter that world and become in time one of the leading figures on the American business scene. But young Henry had different ideas. While still a Princeton undergraduate he became enamored with the study of backboned animals, ancient and modern—so much so that he went with his friends W. B. Scott and Francis Spier to hunt for fossils in the Bridger Basin of Wyoming in 1877, at which time that region was a rather hostile wilderness. Then, overcoming his father's initial objections, to such a degree that the elder Osborn soon wholeheartedly supported his son's ambitions, Henry did graduate studies in England and Germany with full parental support. Thus, by virtue of his upbringing and education, Osborn began his scientific career as a confident, suave Victorian gentleman. He had the virtues and the faults of an upper-class Victorian person, the sort one encounters in the novels of Edith Wharton.

Matthew, on the other hand, grew up in a world of harder knocks, a world with more adversities, where one had always to work hard, where there were no ameneties to be had from the ministrations of servants, and where every penny counted.

Yet Matthew was in no real sense disadvantaged. He grew up with the comfort of a loving extended family and had the close support— by precept and example—of his paleontology-oriented father. His early schooling was of a plebeian sort, but if there were any deficiencies of academic opportunities they were more than compensated for by his being blessed with a mind of remarkable perspicacity. He was ready at thirteen to enter college, but purposely delayed the beginning of his higher education for three years, to be more in step with his classmates. As for his graduate school years, he did not have the experience of studying abroad (considered so desirable in those last years of the nineteenth century), but at Columbia he came under the influence of first-class intellects in geology and biology—including Osborn.

So it was that at the turn of the century these two very different men were intimately associated in their daily work: Osborn, rich and privileged—Matthew, very middle class; Osborn, a believer in class distinctions—Matthew, a true democrat, in spite of being a subject of the Queen; Osborn, colossally arrogant—Matthew, modest but confident of his own abilities; Osborn, domineering—Matthew, independent; Osborn, a man of original ideas often based upon the work of others—Matthew, likewise a man of original ideas mostly based upon his own intimate knowledge of the hard facts of paleontology. Is it any wonder that, even though they respected each other and worked together through the years, inevitable differences should crop up between them?

Matthew, being a forthright person, did not shrink from telling Osborn what he thought, and this did not smooth the common path they traveled. In this respect Osborn was ambivalent; he did not like being disagreed with, yet he did appreciate hearing opinions different from his own. To put it another way, he respected independent thinking on the part of those who worked with him, yet was most comfortable when their thinking followed Osbornian lines. So when Matthew upon occasion quite justifiably differed with "the professor" (as Osborn was known to the residents of the Department of Vertebrate Paleontology at the museum), hackles were raised.

Matthew's colleague, William King Gregory, had the knack of disagreeing with Osborn without ruffling the professor's feathers. He would address the professor in a written memo as "our own imperial mammoth" or "our beloved sulphur-bottom whale" and then proceed to set forth some stand quite at variance with what Osborn had concluded; thus Gregory survived disputes with few if any scars. It was a type of diplomacy that Matthew did not desire, nor was able, to practice.

It should be pointed out that the positions of Matthew and Gregory in relation to Osborn were not the same. They were both Osborn's students—Gregory a few years behind Matthew—but they were students of very different stripes. Osborn "inherited" Matthew as a full-fledged geologist, to be turned around almost 180 degrees, geologically speaking, from hard rock geology to fossils. It was therefore predestined that Matthew would approach fossils from a geological point of view—a position not always understood or appreciated by Osborn. Gregory was already a biologist and, therefore, more attuned than Matthew to the Osbornian comprehension of ancient life. Furthermore, Matthew began his career at the museum as Osborn's junior partner; he worked with Osborn, but not under him. Indeed, within a few years Matthew was established as the person in direct charge of research on fossil vertebrates, although Osborn had the overall supervision of the department. Gregory began as Osborn's assistant and to some degree maintained this relationship throughout their scientific lives together—even after Gregory attained his own independent position at the museum. So perhaps their roles, both actual and perceived, determined to some degree the attitudes that developed between the older man and, respectively, his two younger colleagues.

From the beginning Matthew worked on his own research projects: on Puerco mammals from the San Juan Basin, on early carnivorous mammals, on the Tertiary mammalian fauna of northeastern Colorado, on the classification of fresh-water Tertiary deposits of western North America, to name a few. At the same time he was busily coordinating the activities of several field parties, working simultaneously in separate areas of western North America.

From the beginning Gregory functioned as Osborn's assistant, but he assisted Matthew as well as Osborn. And even though he soon developed his own research interests, particularly in the comparative morphology of fossil and recent vertebrates, he continued to do much of the "donkey work" upon which Osborn based his published contributions. Furthermore, he functioned as editor of some of Osborn's publications, a notable example being a book on the evolution of mammalian molar teeth, published by Macmillan in 1907. Finally, Gregory assisted Osborn in the classroom for several years—until 1910 when he took full charge of the Columbia course.

Matthew's first assignment as a member of the museum staff was to go down into that hot, humid North Carolina mine to collect Triassic reptiles—a thoroughly dirty, disagreeable collecting job. But it was something that had to be done, and Matthew did it without hesitation, and cheerfully. With this initial assignment of his paleontological career we see him undertaking his first field season, the forerunner of many excursions into the countryside in search of fossils. It was an aspect of his paleontological life of which he was very fond.

By way of contrast, Osborn was never a field man. He believed in the importance of fieldwork; indeed he brought Matthew into the museum department in part to carry out the field explorations that were part of the comprehensive paleontological program that he had envisaged in 1891. Yet, as he readily admitted, Osborn did not find fieldwork to his liking. He did sometimes venture into the wilds on an inspection trip, for example, to the Bone Cabin quarry in 1899, but such visits were for the purpose of seeing how work in the field was progressing. He didn't attempt any of the actual work.

Perhaps he was disinclined to do so for good reason. Osborn had come of age in a nineteenth-century moneyed family and had never had to experience physical labor. In his social world other people were paid to dig and delve; such menial work was not a part of his thinking or of his background. A lot of digging and delving is required in the collecting of fossils; these familiar activities would be welcome to Matthew, the independent young man from Canada. But they were beyond the limit for Osborn, the patrician from "Castle Rock" on the Hudson.

So from their very different backgrounds Matthew and Osborn developed a working relationship in laboratory and field—a relationship that at first was harmonious enough, but that, as the years passed, became increasingly strained. In the laboratory—which signifies the quiet hours of paleontological analysis leading to descriptions of fossils, the interpretations of their significance, and the discussions and planning sessions that are recurring necessities—Osborn and Matthew were colleagues, one older than the other, who labored with concerned dedication on their intertwined problems. As Osborn became more self-centered and dogmatic through the years, and as Matthew grew in scientific stature, their differences of opinion regarding paleontological facts and theories became more sharply defined. This was bound to affect their relationship.

In the field the situation was different, because Matthew was the well trained, experienced field man who quite definitely knew what he was doing, while Osborn was the visiting supervisor whose impact on the work being done was of a transitory nature. Osborn was, in a sense, an interested amateur out in the field. Furthermore, he was not out there very much. His concepts of geology, although often forcefully expressed, were likely to be remarkably naive.

I saw a small example of this when I was working as Osborn's research assistant, a few years before his death. He had asked me to review something in the way of stratigraphic analysis (I don't now remember what it was) that was a bit on the unrealistic side. I tried to point this out to him. In a note, set down in his large, bold hand, he wrote "Exactly the same point of view was taken by Dr. Matthew—when I requested him to make with Granger a faunal *life zone survey of the Bridger.* He assured me it would be impossible! HFO." At that point I felt I was in good company.

Something should also be said about Matthew's and Osborn's differences of thought resulting from the divergent bases from which they viewed life, past and present. The pervasive theme of their work was, of course, evolution. No matter how limited particular problems may have been—the description of a skull, the details of tooth anatomy, the listing of sequential strata—such descriptions and listings were always displayed against an evolutionary backdrop.

Evolution for Osborn was a force within the organism, something that determined the direction of development through time. With such a philosophy it was almost inevitable that Osborn would believe in orthogenesis, in rectilinear evolution proceeding from a beginning to a more or less predetermined end. For him the evolution of life in the broad sense, and of particular lines of development in a narrower view, formed a logically beautiful picture. The universe was an orderly place.

For Matthew the picture was not so orderly; indeed it was a bit messy in places. Matthew saw no inner evolutionary drive; rather, he saw animal populations developing as a result of natural selection strongly affected by world climate. Thus the direction of evolution within any particular phylogenetic line might change according to the external circumstances of the environments. "Whereas Osborn and most experimental biologists of the day looked to some factor internal to the organism to explain evolution, Matthew attempted to understand descent with modification in relationship to changes in the climate and elevation of the earth's surface."[1]

Thus Osborn saw the relationships of animals in vertical terms, with bonds of consanguinity stretching through long periods of time, whereas Matthew, although recognizing the validity of lines of descent (as did any paleontologist worth his salt), also saw the relationships between groups of animals of the same geologic age—that is, in horizontal terms.

[1] Ronald Rainger, "Just Before Simpson: William Diller Matthew's Understanding of Evolution," *Proceedings of the American Philosophical Society,* 130 (1986):459–60.

The difference between the thinking of these two men is nicely illustrated (as has been done by Ronald Rainger) by comparing Osborn's phylogency of proboscideans, published posthumously in 1936, with Matthew's phylogeny of horses, published in 1930.

In Osborn's scheme of proboscidean evolution nothing is ancestral to anything—evolution along parallel lines is carried to ridiculous extremes. In Matthew's view of equid evolution there are vertical lines of descent, with ancestral genera giving rise to descendant genera, but there are also horizontal lines of relationship to be seen. The two phylogenies exemplify the contrast between Osborn's concept of evolution proceeding as a result of inner forces within the animal involved, unaffected by individual variability or response to changing environmental conditions, and Matthew's concept of evolution through natural selection.

Osborn believed that new characters in organisms evolved in a regular manner and, being a person who liked to coin terms, he called these changes rectigradations. Later he labeled them aristogenes, since he believed that they were controlled by the genetic nature of ancestral types. He designated his theory of evolution "aristogenesis." It was a term dear to, and widely used, by him, but it did not find favor with other paleontologists or biologists.

It obviously did not impress Matthew, who discussed evolutionary processes in terms of variation and natural selection. In this respect Matthew's view was in the mainstream of evolutionary theory, based upon a Darwinian understanding of organic evolution. He fully accepted the exciting discoveries being made in genetics during the early years of the twentieth century, owing to the rediscovery of Mendelian inheritance theory. He felt that such thinking was in full harmony with the evidence of the fossils—a feeling that was never shared by Osborn.

Indeed, Osborn was quite antagonistic to the geneticists at least as expressed by the work of Thomas Hunt Morgan. This posture was quite the opposite of Matthew's who wrote that "the nature of these variations is much better understood than it was in Darwin's time, thanks especially to the researches of T. H. Morgan and his school."[2]

We see here a basic contradiction in the attitudes of the two paleontologists—a rigidity of thought on the part of Osborn; flexibility on

[2] W. D. Matthew, "The Pattern of Evolution," *Scientific American,* 143 (1930):193.

the part of Matthew. It was inevitable that the philosophical rift between them would so widen that, in the end, collaboration would become impossible.

The extent of their incompatability became apparent on a project having to do with the evolution of the horses—a subject to which Matthew had devoted years of research and thought. There was to be a large monograph on the fossil horses of North America, a joint work by Osborn and Matthew, and to this end Matthew had expended hours studying, analyzing, writing, and supervising the work of artists and the like, not to mention the months spent in the field collecting fossils. Yet, after all of this effort, Matthew finally felt compelled to withdraw from the project, an action that must have been accompanied by bitter disappointment.

So it was that in 1918 there appeared a large format Memoir of the American Museum of Natural History—217 pages, 54 plates, 173 text figures—entitled "Equidae of the Oligocene, Miocene, and Pliocene of North America." The author: Henry Fairfield Osborn.

That monumental volume was to a large extent Matthew's work, and it represented the results of a research project that was particularly dear to him. Even at this distant date it seems extremely unjust that Will Matthew's name is not on the title page of the Memoir. Why could not Osborn have made some adjustments in his attitude that would have preserved joint credit for the authorship of the volume?

Statements made by Robert L. Evander, of the Department of Vertebrate Paleontology at the American Museum, in a recent letter to Dr. Joseph Gregory, of the Department of Paleontology at Berkeley, add information indicating that the monograph on horses was essentially Matthew's work.

October 2, 1991

Dear Dr. Gregory:

I wish to report that I have followed your instructions, and that I have relocated the notes from which Osborn's Iconographic Revision of the Equidae was dictated. Thereby, I have also located Matthew (1913). The document in question is a large book of blank pages, onto which Osborn's thoughts on the various horse species were pasted. The only (evidently) copy of Matthew (1913) was cut into pieces, the various pieces were stamped with a "W.D.M. 1913" stamp, and then the pieces were pasted into the book.

This manuscript documents Stirton's statement that Osborn (1918) is substantially the work of W. D. Matthew. Matthew, for instance, laid out the microevolutionary series *Merychippus isonesus primus* through *quintus* as sp. I through sp. V. Osborn's only contribution to the publication of these names was the Latin ordination. Matthew not only contributed to the description of the various species (for which Osborn gives him some credit), but he also wrote the introductory sections on tooth nomenclature and geologic horizons. Some of the illustrations in Osborn (1918) are simply spruced-up versions of Matthew's sketches. . . .

Sincerely,
Robert L. Evander

Yet, despite these differences in their scientific attitudes Matthew and Osborn worked together in the Department of Vertebrate Paleontology year after year. Probably Matthew's feeling of responsibility to the museum and his fellow workers enabled him to cooperate with Osborn, in spite of scientific disagreements. Indeed it is indicative of Matthew's pragmatic and balanced view of both himself and his work environment that he was able to function for many years as the head of the department and carry on his research with a reasonably tranquil mind.

Matthew did not make a big thing of the difficulties of working with Osborn. Certainly he did not discuss it with his immediate family (with the possible secret exception of Kate), his other relatives, or friends. As mentioned, he did not take museum problems home with him.

But he did make his feelings known now and then in an overt manner. Once Professor Osborn entertained the whole department at his home on the Hudson. On the appointed day the members of the department, great and small, gathered at "Castle Rock" to enjoy the hospitality extended them by the professor. This included a luncheon, and when the time came for one and all to sit down to their repast, it turned out that there were two tables. The one upstairs was for the scientific staff; the other, downstairs in a sort of basement, was for the technicians, secretaries, artists, and the like. When Will Matthew saw the situation, he abandoned the upstairs room, to eat with the plebeian contingent down below. Professor Osborn never quite forgave him for that.

CHAPTER 13

The Triumvirate

*T*he interactions between Matthew and Osborn through the years, important as they may have been to the general well-being and health of the Department of Vertebrate Paleontology, were nevertheless only part of the larger problem of departmental programs and accomplishments that involved all the scientific and support staff. It was with these people that Will Matthew spent much of his time, month in and month out, and it was with many of them that he forged strong bonds of friendship.

It should be emphasized that the American Museum was, and still is, a good place to work. The same holds true for the Department of Vertebrate Paleontology. When Osborn founded the department, he brought together the best talent he could entice to New York, so from the beginning a group of able people devoted to the science with which they were involved gathered there. Of course the best can have their differences, and certainly there were differences within the department. But it would seem that areas of agreement far outnumbered those of disagreement, and that from the outset the department was a pleasant workplace, where people labored together in reasonable harmony.

One of the first to be taken on by Osborn was Jacob Wortman, who had been an assistant to Edward Drinker Cope, the Quaker naturalist, discussed on previous pages of this account. Wortman was a medical doctor by training, but for many years he had been Cope's research associate, in which capacity he had acquired a broad knowledge of fossil vertebrates. An early departmental report noted that "upon June 15th, 1891, J. L. Wortman was chosen Assistant Curator and leader of field work ... and in 1895 W. D. Matthew joined the staff permanently as second assistant." Matthew succeeded to Wortman's position when Wortman was lured to Pittsburgh in 1899 to take charge of vertebrate paleontology at the newly established Carnegie Museum. And as recounted earlier, the Carnegie Museum tried to get Matthew, too, but he resisted the blandishments of that new and richly endowed institution.

Also in the department during its first years were Walter Granger and Barnum Brown (both of whom came aboard in 1896), J. W. Gidley (who joined the group in 1899), and O. P. Hay (who was added in 1900). William King Gregory also became a staff member in 1900. Albert (Bill) Thomson, who figures prominently in the Matthew story, came to New York in 1899 (although he had been working in the field for the museum since 1894) to join the preparation staff of the department. In addition there were Adam Herman, head of the preparation laboratory, Peter Kaisen, one of the early preparators and collectors, A. E. Anderson, the photographer whose technical photographs of fossils have never been surpassed, and the scientific artists, Erwin Christman and Mrs. Lindsay Morris Sterling, whose pen and ink drawings of skulls and bones have likewise never been surpassed.

Another artist with whom Will Matthew collaborated was Charles R. Knight, the gifted depicter of ancient life, whose paintings won for him worldwide fame. Knight was not a regular member of the Department of Vertebrate Paleontology; he was a free-lance artist who spent much of his life working with the departmental staff. He was hired by professor Osborn to make restorations of extinct animals and of the environments in which they lived, works to be done with strict scientific supervision. Although Knight's paintings were done under the direction of Osborn, it was Matthew who supervised the details of these representations of the past. Matthew and Knight had a good relationship; they respected each other and worked together in harmony. Charles Knight often mentioned his intellectual debt to Matthew, expressing his appreciation for the guidance he had received as he labored with brush and paint to make the past come to life. The results were superb; Knight's restorations are still recognized as masterpieces of authenticity—thanks in part to Will Matthew's constructive criticisms.

Matthew's association with Jacob Wortman was of short duration, because Wortman left for Pittsburg four years after Matthew had joined the museum staff. Wortman was a rather enigmatic person— an able scholar and field collector, but a man not easily satisfied. His tenure at the Carnegie Museum lasted about eight years, as had been the case in New York. Then, in 1908 he resigned from the Pittsburgh institution, went to Brownsville, Texas, opened a drug store, and spent the rest of his life as a pharmacist. In the annals of vertebrate paleontology he is one of the few practitioners of the science to have abandoned the field for something quite unrelated to fossils and evolutionary studies. Most vertebrate paleontologists, once established, remain devoted to the subject during their earthly existence.

Of the several persons mentioned there were two, Walter Granger and Albert (Bill) Thomson, who were particularly close to Matthew through the years. Indeed, these three—Matthew, Granger, and Thomson—might be designated the triumvirate. They were about the same age, Matthew having been born on February 19, 1871, Granger on November 7, 1872, and Thomson on February 26, 1874.

Matthew's lifelong friend, Walter Granger, was born in Middletown, Vermont, a hamlet of a few hundred people; he grew up in the comparative metropolis of Rutland, not many miles away. He was a lean, lanky Vermonter, well-acquainted with the rugged hills and the wildlife of that northern state. He would tell stories of life in an environment where winters were long and snow was deep—of how in the late fall his mother would bake dozens upon dozens of mince pies that were stored in a lean-to on the side of the house, where they would freeze solid. Then, on almost every winter morning she would fetch a pie from the lean-to, pop it into the oven of the big coal stove, and serve the family pie for breakfast, New England style.

On September 30, 1890, a telegram for Walter arrived in Rutland from his father, who was in New York on a business trip. Mr. Granger had paid a call on an old acquaintance, Mr. Jenness Richardson, formerly of Rutland, who was the chief taxidermist at the American

Museum of Natural History, and Mr. Richardson had informed Mr. Granger that there was a job available, the next day, at the museum. Walter caught the train that very night for New York and appeared the next day before Mr. Richardson.

It turned out that the job was in the maintenance department and paid the princely sum of twenty dollars a month. Walter took it and found himself assigned the task of cleaning oil lamps that bordered a walkway leading from the street to the museum door. But through the kindness of Mr. Richardson he was also able to work in the taxidermy department, where he spent the first few years of his museum career.

His real interest was in fieldwork and an opportunity for him to satisfy this ambition came in 1894, when he joined a museum paleontological field party led by Wortman. Granger was there to collect birds and mammals while his companions excavated fossils. In the autumn, after his return to New York, he sought the advice of Frank Chapman, the noted ornithologist, who suggested that he might be happier as a paleontological field collector. So the next year he was in the field again with Wortman, this time to assist in collecting fossils as well as birds and mammals. There is a picture extant of this field party, showing Granger and Thomson, both beardless youths, seated behind Wortman—an intent, tough-looking customer with a heavy black beard and moustache and abundant black hair—and O. A. Peterson, briefly a museum paleontologist, but who thereafter spent most of his life at the Carnegie Museum in Pittsburgh.

Obviously Granger's work in the field was satisfactory and appreciated. In 1896 Osborn transferred him to the Department of Vertebrate Paleontology. Thus began his career as a paleontologist, a career that in the beginning was oriented toward collecting fossil vertebrates for research and exhibition. He was involved in the work at the Bone Cabin dinosaur quarry in Wyoming, where giant dinosaurs of late Jurassic age were excavated. Indeed, the triumvirate—Matthew, Granger, and Thomson—was at Bone Cabin in 1899, when the department was engaged in a concerted effort to exploit the dinosaurian riches of that deposit.

Walter Granger continued fieldwork at Bone Cabin where he was in charge of the excavations until 1903. Then his interests shifted to fossil mammals, the branch of paleontology to which he would devote the rest of his life. Thus began the long collaboration of Matthew and Granger, both in the field and the laboratory. A by-product of the collaboration was a friendship that endured through more than three decades—until Matthew's untimely death in 1930.

Walter Granger married his cousin, Anna Dean Granger of Brooklyn, on April 7, 1904, the year before Will Matthew's marriage to Kate Lee of Brooklyn. But there was not a great deal of socializing between the Matthew and Granger families. They were friendly enough, but their family life-styles were so different as to preclude frequent visits, back and forth. The Grangers always lived in a Manhattan apartment, and they had no children. The Matthews had Elizabeth a little more than a year after their marriage and soon migrated up the Hudson to Hastings. The Matthews' evenings, weekends, holidays were very occupied with family, house, and land, so little time was left for getting together with Walter and Anna Granger, fifteen miles away. One summer the Grangers did take over the Hastings house when the Matthews were away. When the Matthews returned, they found a line of squirrel tails nailed to the wall on the back porch. Walter had been more than a little annoyed at the manner in which the squirrels raided the garden, so he took care of the matter with a small rifle.

From 1903 until 1918, with the exception of 1907, Walter Granger collected Paleocene and Eocene fossil mammals, mainly in Wyoming and particularly in the Bridger Basin, and his collections were the basis of studies by Matthew and Granger, both jointly and separately, that constitute a classic series of papers dealing with some of the very ancient mammals of North America. In 1907 Granger went to the Fayum region of Egypt with Professor Osborn, where a choice collection of Eocene mammals was obtained. The collection was made by Granger and George Olsen, one of the American Museum preparators, aided by Egyptians. Osborn, who had his family along, soon took off with the family for other regions—the usual Osbornian fashion.

In 1921 Walter Granger was appointed chief paleontologist and second-in-command of the famous Central Asiatic Expeditions of the American Museum of Natural History, organized and led by Roy Chapman Andrews. That was before the days of large government grants for such work, so the mounting of the elaborate expeditions to Mongolia, which aroused worldwide interest, was no mean accomplishment by Andrews.

Granger pursued much of his fieldwork apart from Matthew, although they collaborated extensively in the study of fossil mammals that Granger collected, and Matthew worked occasionally in the Bridger Basin.

Although Walter Granger had no formal university training, he became a paleontological scholar of distinction through his association with Matthew—so much so that he was awarded an honorary doctorate in 1932 by Middlebury College, in Middlebury, Vermont. As an indication of Walter Granger's level of attainment in paleontological scholarship, the series of Matthew and Granger papers on Wasatch and Wind River fossil mammals from Wyoming may be cited. There were nine contributions, four by Matthew and Granger, four by Granger alone, and one by Matthew.

These two members of the triumvirate stand tall in the annals of mammalian paleontology. Theirs was a partnership to which each member contributed his own special talents. George Simpson wrote of the Matthew-Granger collaboration that "he [Granger] contributed to these [joint studies] not only the specimens and the field data but also a soundness of judgment and acuteness of perception that were, as Matthew frequently remarked, essential to the scientific value of the results. . . . His interest in all such studies was keen and his untiring, unselfish assistance was endless and practical and could be acknowledged only over his protests."[1]

Albert Thomson, the youngest member of the triumvirate, was born on February 26, 1874, in Elk Point, South Dakota, which is in the southeastern tip of that state where a narrow point thrusts down between Nebraska and Iowa. But Albert, known for some reason as "Bill", was truly a westerner. He grew up on his father's ranch near Rapid City and not far from the Black Hills, where he learned all of the skills of an old-time cowboy. The family had a house in Rapid City, and there Bill attended school.

In the summer of 1894, when he was twenty years old, he joined Wortman's American Museum expedition as cook, teamster, photographer, and general handy man. His services were so appreciated that he spent the following five field seasons with American Museum parties, including three years at the Jurassic Bone Cabin quarry in Wyo-

[1] G. G. Simpson, *Science,* 94 (1941):339.

ming. In 1899 he went to New York to join the Department of Vertebrate Paleontology, and with that department he spent the rest of his paleontological career. He was the first member of the triumvirate to abandon bachelorhood; he was married to Mary Gildea on April 2, 1902.

Will Matthew and Bill Thomson became field partners early in their careers. They were together as part of the large group that worked in the Bone Cabin quarry in 1899, and after that Thomson's collecting activities were entirely devoted to Cenozoic mammals, which meant a close association with Matthew for twenty years.

During the first few years of the new century Bill spent his summers in home territory, in the Oligocene and Miocene sediments of Nebraska and South Dakota as a participant in Osborn's strategy of simultaneous field exploration by several parties, collecting late Mesozoic reptiles and Tertiary mammals from the western states. The purpose was to build up extensive collections as the basis for studies of vertebrate evolution as well as the succession of vertebrate faunas in this part of the world. The program was eminently successful; excellent beginnings were made toward the amassing of comprehensive collections of fossils that were to establish the American Museum's worldwide preeminence in vertebrate paleontology.

In the summer of 1906 Bill joined Will Matthew for a paleontological campaign in South Dakota, of which we have a detailed record thanks to Matthew's almost daily letters to Kate, portions of which are set forth in part, in chapter 10. Their work together in 1906 established a pattern of close cooperation and mutual trust that was to prevail for two decades. It was a pattern whereby Matthew could spend only a part of the season in the field with Thomson, confidently leaving to that triumvirate partner the responsibility of carrying on the work, when other duties—inevitable for a person in Matthew's position—required a departure from the field. The strategy is adumbrated in two of Will's 1906 letters to Kate.

I told you of Osborn's wanting to come out. I would like very much to go over the ground with him, but it isn't possible for this year at all events. [Matthew had to go back to New Brunswick for the birth of Elizabeth.] I hope he'll come out any way and that Thomson can take him 'round and show him how things are. I think that there is every prospect now that he will continue to find fossils until the end of the season and by then will have a good collection. Osborn endorses our plans for systematic collecting in this formation under Thomson's charge, which is also satisfactory.

(letter of July 20, 1906)

Also we have first-class prospects for continuing the work along this line for the rest of this season and for several seasons to come. This also is a relief as I will not have to plan and decide on what field to work in (per *Thomson*) next summer.

(letter of July 24, 1906)

In 1908 there began the work of Matthew and Thomson, but especially Thomson, at Agate, Nebraska, after which were the several years of work at Sheep Creek and Snake Creek, Nebraska, that formed the basis for a series of important paleontological papers by Will Matthew.

The fieldwork in Nebraska, at Agate, Sheep Creek, and Snake Creek, occupied Bill Thomson's summers (with Matthew's periodic attendance) through 1927. That year Matthew left New York for California, as shall be related in due course, which resulted in an interruption of the close fellowship of the triumvirate. Of course Matthew, Granger, and Thomson kept in touch with each other, by letter, and in person when Matthew came to New York in the summertime to continue some of his research projects. In 1930, as will be discussed in detail later, when Granger and Thomson were in Mongolia on one of the Central Asiatic Expeditions led by Roy Andrews, Matthew died, suddenly and unexpectedly, and the triumvirate suffered its first loss—thus ended more than three decades of fellowship and shared experiences in field and laboratory. Granger and Thomson quite naturally were grieved by the death of their old friend and felt his absence during their remaining years. But they carried on their work at the museum—Granger in his office looking out across Central Park, Thomson a few steps away across the hall in his laboratory.

It was comforting for them to have each other as companions during those later years. I can remember on more than one occasion Granger walking from his office across to Bill's laboratory with a broken fossil, and giving it to his friend to repair. "It came apart in me hands," Granger would say. "You old devil," Bill would invariably reply. And, somehow, much love was expressed in that ritual exchange.

In his last years of association with the museum Bill Thomson was back in his beloved Dakota country each summer, collecting Oligocene mammals. In the summer of 1941 Granger was with him, as was I. This was a new triumvirate of sorts, with myself as the third, very junior member. One night during this field trip Walter Granger died in his sleep—a year or so short of his seventieth birthday. It was a wrenchingly sad occasion for Bill and me; we closed down the fieldwork and returned to New York, carrying Walter Granger's ashes with us.

Bill, the remaining member of the old triumvirate, stayed on at the museum until the end of 1941 and then retired to live a lonely life in a Bronx apartment. (His wife had died a few years previously.) I used to visit him, to sit and talk, and to listen to his tales of Matthew and Granger. Bill, the last member of the triumvirate, died in 1948.

CHAPTER 14

The World of Early Mammals

At an early stage in his paleontological career Will Matthew became involved with ancient mammals—those warm-blooded, furry, backboned animals that inherited the world after the demise of the dinosaurs—and this involvement occupied a considerable part of his research time for the rest of his days. It began in part with his work on the Cope Collection, which contained many fossil mammals excavated from early Cenozoic sedimentary rocks in the western states, and in part because of his collaboration with Jacob Wortman on such fossils collected by Wortman and his associates, and in some measure (one would suspect) because of his unceasing curiosity. Will was attracted to problems that challenged his paleontological acumen. Certainly the many aspects of mammalian evolution during those initial millennia of Cenozoic history, when the primitive mammals were establishing their hegemony across continents very different from those of the modern world, were sufficiently challenging for this young paleontologist.

What were the early mammals that so intrigued Matthew that his first paleontological research was devoted to them, thus setting a pattern of inquiry that was to continue throughout his life?

During the 150 million years of the Mesozoic era when the dinosaurs were dominant on land, the mammals were, in essence, suppressed by the reptiles. It is an often unappreciated fact that the first mammals and the first dinosaurs made their appearances on the earth at approximately the same time, in the latter part of the Triassic period. Whereas many of the dinosaurs rapidly—in an evolutionary sense—developed into giants, the mammals throughout late Triassic, Jurassic, and Cretaceous years remained small, "insignificant" creatures, inhabiting hidden, crepuscular environments where they found protection from various aggressive Mesozoic reptiles. This was their role until the transition from Mesozoic to Cenozoic time, when the dinosaurs and some of their reptilian contemporaries became extinct, thereby vacating broad continental regions and making them available for colonization by other creatures. These creatures were the mammals.

The mammals living at the end of the Cretaceous period were the monotremes (represented today by the Australian duck-billed platypus and spiny echidna), the marsupials (represented today by the pouched mammals of North and South America and Australia), the placental insectivores (represented today by such animals as moles and hedgehogs), and a few other placental groups long extinct. These were the successors of the dinosaurs.

The earliest Cenozoic mammals, those living in the Paleocene epoch, were generally small or of moderate size. The Paleocene has been called the epoch of conquest, but these early conquerors were anything but impressive. The Puerco fauna of New Mexico—the first mammalian assemblage to be studied by Matthew—was made up of such animals. They were for the most part rather similar, but a dichotomy of plant-eaters and meat-eaters can be recognized. The most common of the plant-eaters were the condylarths—rather clumsy looking and not adapted for rapid running. The most common of the meat-eaters were the creodonts—primitive mammals that looked something like small, heavy dogs. There were marsupials and insectivores, descended from Cretaceous ancestors that had lived with the dinosaurs, as well as some interesting Cretaceous hold-overs known as multituberculates, rodentlike mammals that were numerous at the beginning of Cenozoic history, until they were finally crowded out by the true rodents.

Interestingly, the first primates, the ultimate ancestors of our own line of evolutionary development, also made their appearance in Paleocene time, the descendants of insectivore progenitors.

From such Paleocene mammals the Eocene faunas evolved. Here are found the beginnings of modern evolutionary lines, with numerous hoofed mammals, including ancestral odd-toed horses, tapirs, and rhinoceroses, as well as ancestral even-toed hoofed mammals, the predecessors of such groups as pigs, camels, and deer. In addition, there were certain hoofed mammals known as uintatheres, the giants of the Eocene world. These rather weird herbivores, as large as modern rhinoceroses, had three pairs of horns on the skull, and large, daggerlike canine teeth. The primates, some of them very like modern lemurs, were numerous and varied. And the rodents, eventually to become the most numerous of all mammals, were well established. There were bats in the air (and birds as well, enjoying a heritage that extended well back into Mesozoic times. (One interesting feature of the Eocene faunas was the presence of gigantic, flightless birds, typified by *Diatryma,* described by Matthew and Granger from the Eocene beds of Wyoming. This huge predatory bird, well adapted to run rapidly on its long, strong legs, was a dominant carnivore of its time. Yet it did not establish a strong trend in evolutionary history; the carnivores were to be predominantly mammals.) Mammals were supreme, and the Eocene mammals were well on the way toward becoming the mammalian communities familiar to us.

The Paleocene, Eocene, and Oligocene epochs, which collectively may be designated as the Paleogene interval of Cenozoic history, were periods of lowlands and tropical climates. It was this early segment of Cenozoic history to which Matthew first turned his attention.

His Paleogene studies, which continued throughout his life, may be divided into three phases. One was concerned with the Paleocene sediments and faunas of the San Juan Basin of New Mexico. Another was directed toward the Lower Eocene sediments and faunas of the Big Horn Basin of Wyoming. The third involved the Middle Eocene sediments and faunas of the Bridger Basin of Wyoming.

The San Juan Basin of New Mexico was first paleontologically explored by David Baldwin, a free-lance collector who lived in Abiquiu, New Mexico, and who spent much of his time in the 1880s prospecting and collecting for Edward Drinker Cope. He discovered the Puerco fauna, the oldest Paleocene assemblage, and the fossils that he had collected were described by Cope. They were part of the Cope Collection, purchased by the American Museum of Natural History, the collection that Will Matthew packed and catalogued in 1897. In 1892 and again in 1896 Jacob Wortman explored the San Juan Basin, adding many new fossils to the collections that had been accumulated by Baldwin. These Baldwin and Wortman collections were the fossils on which Matthew truly began his paleontological career.

The foundation stone of that career was a sixty-four page paper published in the *Bulletin of the American Museum of Natural History*, entitled "A Revision of the Puerco Fauna." It was his first extensive paleontological contribution and a harbinger of the significant studies that were to come. In this paper he affirms the presence of two Paleocene faunas in the San Juan Basin, as first suggested by Wortman—a lower, Puerco, fauna and an upper (and later), Torrejon, fauna. He carefully revised the genera and species that constituted the two faunas, showing how these primitive mammals in many cases retained relationships with each other, even though they were evolving along lines that were to become increasingly differentiated. For example, the primitive hoofed mammals of the Puerco and Torrejon faunas, the condylarths, were not far removed from the primitive carnivorous creodonts. But in later geologic time the descendants of these evolutionary progenitors were to become quite different in their specializations, as we can see today when we compare a wolf with a horse.

From this first paper there grew a program of research that was to continue through Matthew's lifetime, culminating in the huge monograph on the Paleocene faunas of the San Juan Basin, published by the American Philosophical Society in 1937, seven years after his death. One wishes he could have lived to see this great work in print.

As recounted earlier, Walter Granger moved in 1903 from Bone Cabin Quarry, where much of his energies had been devoted to the heavy work of collecting giant dinosaurs, to the Bridger Basin in the southwestern part of Wyoming, where, for the better part of fifteen field seasons, he would be involved with early mammals, most of them quite small, compared with the dinosaurian giants. Those field seasons were not, however, devoted solely to work in the Bridger Basin, because, beginning in 1909, Granger divided his time between the middle Eocene sediments of the Bridger and the lower Eocene beds of the Wind River and Bighorn Basins that lie between the Wind

River and Bighorn mountain ranges, over two hundred miles to the north of the Bridger Basin.

Matthew joined Granger in the Bridger Basin in 1904, but for the most part these field projects were carried on by Granger and his assistants while Matthew was occupied with other duties. Nevertheless, the Bridger Basin project—involving fossils from the Bridger Formation—and the Wind River–Bighorn Basin project—involving fossils from the Wind River and Wasatch Formations—respectively, were classic examples of cooperative field work—laboratory preparation—detailed research—publication, in which Matthew and Granger worked together through many years.

Fieldwork in the Bridger Basin had been carried on by several paleontologists before Granger entered this area in 1903. The first fossils were collected by the famous Hayden Survey, soon after the Civil War, and were described by Joseph Leidy, the paleontological sage of Philadelphia in 1858–70. As might be expected, those two implacable antagonists, Marsh and Cope, entered the Bridger Basin during the decade of the seventies—Marsh characteristically depending on the efforts of his hard-worked assistants, Cope becoming personally involved as paleontologist of the Hayden Survey. Both men published papers describing fossils that had been discovered. Also it should be pointed out that the Princeton Expedition of 1877—on which Osborn and Scott, then very young men, experienced their baptismal collecting experiences—was centered in the Bridger Basin. And finally, Jacob Wortman collected there most successfully in 1893, after which, in 1901–3, he described some of the Bridger fossils in the American Museum collections. Thus when Granger began his Bridger campaign he was not venturing into an unknown land; there was a solid base on which he was to build his own work.

This he did, so that with the materials Granger added to the previous collections housed at the American Museum, Matthew began his Bridger studies with a wealth of fossils at his command. The result was the magnificent large format memoir entitled *The Carnivora and Insectivora of the Bridger Basin, Middle Eocene*. This was a lengthy publication, the likes of which are seldom seen in today's age of publishing restrictions. The 276 oversize pages made this one of Matthew's two largest publications, and its appearance gained him worldwide recognition.

This was the work on which Will was working so hard when he

and Kate were newly married. It was the publication that she duti-
fully felt she had to read and that, with expressions to her husband of
regret, she soon abandoned: it was hardly intended as a fireside book.

As the title indicates, this was a revision of the description of primitive insectivores and carnivores that lived during the early years of
mammalian evolution, and a most thorough and penetrating revision
it was. But it was more than a reconsideration of two groups of mammals. Matthew also described the Bridger Formation in its various
aspects, how the sediments were deposited, the divisions within the
formation, the succession of fossils found there, and the ecology of the
fauna. It was a picture of life fifty million years ago in what is now
Wyoming.

This assemblage of mammals contained archaic elements, such as
the ancient carnivores known as creodonts—predators that to modern
observers seem to have been just a bit clumsy, and the huge six-horned
uintatheres already mentioned. But many of the Bridger animals had
something of a modern look. There were the basic ancestors of the
modern carnivores, as well as lemurlike primates, insectivores related
to modern moles and hedgehogs, various rodents, odd-toed hoofed
mammals represented by early horses, tapirs, rhinoceroses and
rhinoceros-like titanotheres, and even-toed hoofed mammals in their
primitive manifestations.

It was a picture of life greatly advanced over that of the Puerco-
Torrejon mammals that Matthew described based on the finds in the
San Juan Basin of New Mexico.

While Matthew was minutely studying the fossils from the Bridger
Basin, and writing the manuscript of his great monograph, Walter
Granger was beginning his collecting activities in Wyoming—which
not only would add more Bridger materials with which Matthew
could augment his research, but also began the accumulation of
Lower Eocene fossils from the Wasatch and Wind River Formations
of the Bighorn Basin. On this long-term project Granger was more
than the collector and adviser who stood in the background, to some
degree overshadowed by his colleague who was studying the fossils
and writing descriptions and analyses for publication. From the beginning the Wasatch–Wind River studies were truly a collaboration
between these two old friends.

Granger and his assistants did the collecting, while he and Matthew supervised the preparation of fossils in the laboratory between

field seasons. Then they allocated the research between themselves and published a series of American Museum Bulletins that described the various groups of mammals that comprised the Wasatch and Wind River faunas.

These mammalian faunas of early Eoccne age were, as might be expected, more or less intermediate between the very primitive faunas of the San Juan Basin and the later Eocene Bridger fauna that Matthew had described and analyzed in his monograph of 1909. They comprised archaic elements, inherited from the Paleocene Puerco and Torrejon assemblages, and progressive mammals that were becoming increasingly numerous in the Bridger fauna. Indeed, as Matthew pointed out, the Wasatch fauna consists of a nearly equal mingling of archaic and "modernized" mammals—the latter being ancestors of the mammals that later were to inhabit the forests and prairies of North America.

As for the archaic elements in the Wasatch and Wind River faunas, one has only to look at those hoofed mammals that may conveniently be designated as condylarths and amblypods—the former typified by the genius *Phenacodus,* the latter by the *Coryphodon,* both descended from Puerco and Torrejon ancestral forms. *Phenacodus,* about the size of a sheep, was indeed a sort of archetypical ungulate (or hoofed) mammal in which the feet were short with all of the toes present, each toe terminating in its own little hoof. The limbs likewise were rather short, the body was rather long as was the tail, while the skull was low. The jaws had low-crowned grinding teeth, evidently more suited for browsing on soft leaves than for grazing. (Grazing was a later development in mammalian evolution as an adaptation for crushing the hard, siliceous grasses that had not as yet evolved when *Phenacodus* frequented early hardwood forests.)

Coryphodon was perhaps a somewhat more advanced herbivore than *Phenacodus,* showing the increase in size that has so commonly marked evolution among plant-eating mammals. This large, heavy browser, the size of a modern tapir, was obviously a member of the hoofed mammal community, characterized as it was by a barrel-like body, a short tail, and a heavy skull with broad, crushing teeth in the sides of the jaws. But in spite of such herbivorous adaptations, *Coryphodon* also had a pair of long upper, saberlike tusks, sufficiently large to be dangerous in combat.

Looking at the modernized herbivores of Wasatch time, we see that the ancestral horse, *Eohippus* (unfortunately more properly designated as *Hyracotherium,* owing to the implacable international rules of zoological nomenclature), accounts for about one-third of the fossils characteristic of the Lower Eocene Wasatch sediments. Among the hoofed mammals there were others that, although occurring as ancestral types, nonetheless gave these early Eocene associations a protomodern cast. For example, there were ancestral tapirs and rhinoceroses, anatomically not far removed from *Eohippus,* yet definitely occupying the lower rungs of evolutionary ladders that were to lead to the very different odd-toed hoofed mammals familiar in our modern world. And there were the beginnings of the even-toed hoofed mammals as well, the progenitors of the modern ungulates.

The balance of predator and prey that has persisted through millions of years of evolutionary history was maintained in these early Eocene faunas by the presence of carnivores, both archaic and advanced. The rather clumsy creodonts, so prevalent in the basic faunas of the San Juan Basin, were still present, but they shared the scene with rather slender, light-limbed carnivores known as miacids—these being the ultimate ancestors of the carnivorous mammals familiar to us. Perhaps the large, heavy creodonts preyed upon such archaic herbivores as *Phenacodus,* while the miacids directed their attentions toward the more advanced herbivores such as *Eohippus*—small and swift of foot, but easier to pull down when once overtaken.

Insectivores as well as rodents and lemurlike primates were also found in the Wasatch and Wind River faunas.

All these early mammals were the subjects for separate descriptions and analyses by Matthew and Granger. As explained in the introduction to their first paper, "This series of contributions deals therefore with practically all that is known to science of the Lower Eocene mammalia. [That was in 1915.] The authors, while in entire accord as to their conclusions, are separately responsible for the sections of the revision appearing under their individual names, and it is requested that they be so quoted."[1]

Perhaps the foregoing remarks will give some idea of the triple

[1.] W. D. Matthew and Walter Granger, "A Revision of the Lower Eocene Wasatch and Wind River Faunas: Part I—Order Ferae (Carnivora), Suborder Creodonta," by W. D. Matthew, *Bulletin of the American Museum of Natural History,* 34 (1915):1–103.

series of studies, simultaneously conducted through the years, that made Matthew such a profound scholar of and authority on the early mammals. It should be emphasized that although this chapter has focused on Matthew's three largest and most extensive early Cenozoic research studies, his contributions to our knowledge of Paleocene and Eocene mammals comprise some fifty or so monographs and lesser papers (including Mongolian as well as North American studies) published throughout his lifetime. Thus it can be said that the early evolution of the mammals—the phase of their collective life histories that had to do with the conquest of continents recently divested of the ubiquitous, dominant dinosaurs, with the early "experiments" in mammalian adaptation, and with the establishment of the lines of mammalian evolution destined to survive and flourish through millions of years of Cenozoic history—that early evolution occupied his thoughts from the early years when he was "writing hard" as Kate put it, to the day of his death, his huge San Juan Basin monograph still unpublished. Yet in spite of what might seem an overwhelming intellectual commitment, one that would have been quite enough for a lesser scholar, Will Matthew was equally involved in the collecting and study of the later mammals—those of the Miocene and Pliocene epochs that collectively comprise the Neogene interval of Cenozoic history.

CHAPTER 15

Agate and Snake Creek

Although Will Matthew was from the beginning a student of the earliest and most primitive mammal faunas, he was also busy with mammals of modern aspects. As related in Chapter 6, in 1898 he was out in the plains of northeastern Colorado, "stewing" in his tent, resting from the hot and arduous task of collecting fossil animals of Oligocene and Miocene age, such as squirrels, mice, beavers, camels, horses, and various carnivores, including saber-toothed cats. This was the collection upon which he based his first "big" publication—a Memoir of the American Museum, published in 1901 in a large format with pages 11 × 14 inches, ninety-two pages of text plus plates, entitled *Fossil Mammals of the Tertiary of Northeastern Colorado.* It was this collection that so appealed to his aesthetic senses—fossilized bones of such perfect preservation that they had all the appearances of recent bones bleached for exhibition. "I have never seen any quite so perfect" was his remark.

Although these excellent fossil bones formed the basis for his memoir, he turned his attention in the opening pages to the questions of how the sediments of the high plains—from the highly eroded, scenic badlands of the White River of South Dakota (now Badlands National Park) southward to the cliffs and arroyos of northeastern Colorado—might have been formed.

There was a long-standing theory, originated in part by Clarence King—the first director of the United States Geological Survey, who had worked in the western states during the decades after the Civil War—that propounded that the sedimentary rocks of the high plains had been deposited in vast inland lakes, and in these putative lake beds the bones of Oligocene and Miocene animals had been deposited during the eons of the middle Tertiary period. This theory was being questioned by various geologists during the final years of the nine-teenth century, and Matthew, from his paleontological viewpoint re-garded it with considerable doubt.

Consequently, in his monograph he viewed the inland lake theory from all angles, with the evidence of the layering of the sediments and the fossils each contained on which to base his arguments, and then he nicely demolished the theory of King et al. with a series of closely reasoned analyses. As a result of his study of the rocks and fossils of the Oligocene White River badlands and the Miocene "Loup Fork" beds (as they were known at the time) he demonstrated the fol-lowing:

1. If the sediments had been deposited in a vast lake there should have been heavy terraces developed around its shores. No such terraces are to be seen.
2. There is no sedimentary evidence of a lake basin on the Great Plains. The sediments are of the variegated nature of those laid down by shifting streams.
3. There are intercalcations of fine clays and coarse sandstones such as might be expected in the development of river deposits.
4. If these deposits were of lacustrine origin, they should be thickest in the marginal areas and thinnest, or even absent, in the central areas of the supposed great lakes. Exactly the opposite is the case.
5. The contacts between these high plain sediments and the rocks beneath them are of the random nature that would be developed by swiftly flowing streams. They are not graded, as would be the case by the sorting action of an advancing lake margin.
6. The prominent banding of the White River clays, which were thought by earlier students to indicate lake sediments, were just as likely to have been the result of changes in climate.

He then went on to analyze the fossils and showed that in Oligo-cene White River sediments, for example, the fine clays contained fossils of animals that lived on plains or savannahs—rodents, early rabbits, various hoofed mammals, and the carnivores that preyed upon them, while the coarse sandstones contained fossils of animals that frequented river courses—early beavers, various carnivores, ancient horses, tapirs and rhinoceroses (the latter being animals that were most definitely adapted to live a hippopotamus-like existence), as well as early piglike mammals, sheeplike oreodonts and anthracotheres (the true ancestors of the Old World hippos). Thus the faunas of the clays and the sandstones were contemporaneous but contained animals that were in different habitats.

Among the fossils that he discovered in Colorado was *Alticamelus,* which he named, a huge, long-necked camel as tall as a giraffe— hence the name. In addition to the important descriptions in this memoir are numerous excellent illustrations, including his restorations of an early dog, *Cynodictis,* and an early saber-toothed cat, *Dinictis,* which, thanks to his collaborative work with one of the departmental artists (either Lindsay Sterling or Erwin Christman), are both scientifically and esthetically superb.

From the publication of the memoir (truly a tour de force for a young paleontologist at the beginning of his career) through the rest of his life there was a mix in his bibliography of papers devoted to the most primitive mammals and papers having to do with mammals from all levels of Cenozoic history. His comprehensive studies on mammals of all ages, up through the Pleistocene (or ice-age) epoch into modern times, were correlated with his collecting trips in the field, in order that he might gain firsthand knowledge of stratigraphic relationships crucial to his understanding of the fossils he was seeking, while at the same time garnering the very specimens upon which to base his studies. Consequently, he was frequently out in the high plains and badlands, not only to work with Granger in the sediments that yielded the most ancient of mammalian faunas but also to work with Thomson in beds of intermediate and late stages within the Age of Mammals, wherein are found the direct ancestors of modern mammals.

As noted, in 1906 Matthew was with Thomson in South Dakota, collecting mammals, enduring the summer heat on Porcupine Creek, and writing daily letters to Kate. A logical sequel to the collecting activities of 1906 in northeastern Colorado and along Porcupine

Creek in South Dakota, was to be a campaign at Agate and vicinity in the northwestern corner of Nebraska, in 1908.

Why Agate? Because Agate is and has been, for almost a century, a special place for North American vertebrate paleontologists, particularly those bone hunters interested in mammals that lived a mere fifteen million years ago and less. That age may seem of great antiquity to the casual reader, but to geologists such an interval of earth history is like a day. At Agate, and in the surrounding high country of northwestern Nebraska are sandy rocks that contain great quantities of fossil bones—the remains of animals that lived on our continent when the West was probably not so high, so arid, or so cold in the winter as is presently the case.

Agate is a lovely oasis, a green island in the high, sun-baked plains, where grass-covered rugged hills form an introduction to the high Rockies, fifty miles or so to the west. It is an island of lofty trees, among which wind the headwaters of the Niobrara River, at the beginning of its three hundred mile journey eastward to join the great, muddy Missouri. In the midst of this little park (for it is not very large) there is a rather plain, square, white house, bordered by the nascent river and shaded by tall cottonwoods.

Agate, on the upper Niobrara, with its grove of trees and its wide expanses of lawns, is the creation of Captain James H. Cook, one of the remarkable men of western North American history.[1] Captain Cook, cowboy, hunter, guide, scout, and ranchman, acquired this land in 1887 and brought to it his young wife and his baby boy, Harold. He built the house and planted the trees where no trees had grown and nurtured them during the crucial early years of growth. Harold Cook wrote, "I can remember seeing him after he had come in from riding or working in the corrals, dog-tired and weary, carrying buckets of water to each tree, individually, hundreds of them, to keep the seedlings alive and growing until their roots could push down to water."[2] His labors were rewarded.

---

[1] For anyone wishing to learn something about the West during the latter decades of the nineteenth century, a book by Captain Cook is strongly recommended. Here is a first-person account of high adventure by a perceptive man, who participated in some of the final aspects of the Caucasian conquest of western North America. It is a story that rings true—unlike most western novels and almost all western cinema. See James H. Cook, *Fifty Years on the Old Frontier*. Yale University Press, 1923; University of Oklahoma Press, 1957.

[2] H. J. Cook, *Tales of the 04 Ranch* (Lincoln: University of Nebraska Press, 1968), 12.

About four miles to the east and down the river from the ranch house two buttes rise out of the grassland. One is more or less conical in shape, the other is larger and has a flat top. In 1886 James Cook had brought his fiancée here on a horseback ride, and at the base of the larger hill they noticed fragments of bones scattered on the ground. When they picked up these bone fragments they found them to be heavy, the "marrow cavity" of each filled with glittering calcite crystals. James Cook realized he had found fossils, and eventually he made his discovery known to men who were interested in such things. As a result, in 1891 and 1892 Professor Erwin H. Barbour of the University of Nebraska made the first tentative collections of fossils, obtained simply by picking up specimens from the surface of the ground. Soon after the turn of the century collecting at the two hills began in earnest. In 1904 O. A. Peterson of the Carnegie Museum in Pittsburgh discovered a fabulous "bone bed" on the southwest side of the larger hill, and there he spent the summer extracting fossil bones and blocks of bones from a tangled fossiliferous mass of unbelievable complexity. These were for the most part the bones of a little rhinoceros, which stood about three feet high at the shoulder, and was characterized by a unique pair of horns on the nose, situated side by side, rather than in the fore and aft arrangement seen in other rhinoceroses, both fossil and modern. The concentration of fossils in this deposit was very high; careful studies have revealed a concentration of specimens in parts of the quarry amounting to a maximum of forty bones per *square foot,* which means that the bones are piled one on top of another like jackstraws. Of course museum representatives were attracted to the site of such a paleontological bonanza like miners to a rich vein of gold.

In 1907, at a small hill about two miles to the east by south of the two Agate hills, Professor Frederick B. Loomis of Amherst College and his accompanying field crew discovered a fossil deposit consisting not of a profusion of bones as at Agate, but of scattered skeletons and bones, all belonging to a little camel, which Loomis described and named *Stenomylus hitchcocki.* This graceful, long-legged camel, more like a small antelope than a modern camel in its proportions, stood about two feet high at the shoulder—its skeleton composed of remarkably slender bones. The hill from which *Stenomylus* was excavated was quite logically dubbed Stenomylus Hill, and through those early years of exploration and excavation it was worked by various museums.

The larger of the two Agate hills was named Carnegie Hill by Professor Barbour for the very obvious reason that the Carnegie Museum's Peterson and crew had discovered and were digging out the incredible deposit of fossil bones.

Barbour and a crew from the University of Nebraska began in 1904 to excavate in the other Agate hill, so he called it University Hill. The names, so informally given to the two landmark hills, immediately caught on and continue to this day as the designations for the buttes. A much smaller hill, to the southeast of Carnegie Hill, was named Beardog Hill in 1984 by Dr. Robert M. Hunt, Jr., of the University of Nebraska, who in recent years has intensively studied the Agate fossil deposits and has published a detailed interpretation of them.[3] To the north of Carnegie and University hills is a low ridge, where a Carnegie group briefly worked a quarry—Quarry A. This feature was designated North Ridge by Dr. Hunt.

Hunt has shown that the deposits traversing the two hills at Agate were formed in a stream channel that occupied this area during the early part of Miocene time, that is to say about fifteen million years ago. The land was not so high as it is today, and was probably semiarid with an alternation of wet and dry seasons with a climatic regimen warmer than today. The carcasses of various animals, particularly rather large mammals, were deposited along this stream during alternations of wet and dry cycles, and, depending on local conditions, the bones of these animals were accumulated in varying degrees of concentration. Along the stream banks the concentration of bones might have been rather moderate; in pools or backwaters disassociated skeletons piled up in extraordinary concentrations. That was the case with the rhinoceros bones that so astonished Peterson when he first opened the Carnegie quarry—now called the Southwest Excavation.

The rhinoceroses were for many years thought to belong to a single genus, *Diceratherium,* so named because of the two horns, side by side, on the nose. It is now evident that there were two rhinoceros genera

[3] Robert M. Hunt, Jr., "The Agate Hills: History of Paleontological Excavations, 1904–1925," *Review of Paleontological Resources at Agate Fossil Beds National Monument, Sioux County, Nebraska* (Prepared for the Midwest Region, National Park Service, Omaha, Nebraska. 1984.)

at Agate. The other form—designated *Menoceras*—differed in various anatomical details from *Diceratherium.*

Two other large mammals can be found in the hills at Agate, a perissodactyl or odd-toed hoofed animal (as are rhinoceroses) known as *Moropus,* which was a large horselike herbivore with three clawed toes on each foot instead of hooves, and a huge entelodont named *Dinohyus,* a piglike animal as large as a bison, with an inordinately massive head and two hooves on each foot—the cloven-hoofed condition typical of the artiodactyls or even-toed hoofed mammals.

Moropus, the clawed perissodactyl, proved to be rather rare at Agate and is known today from a couple of dozen skeletons, seventeen of which were excavated by Bill Thomson between 1912 and 1916. *Dinohyus* is even more rare, represented at the present time by only two full skeletons, one collected at University Hill by the University of Nebraska, the other found at Carnegie Hill by the Carnegie group.

Of course this summary represents the accretion of knowledge built up through years of digging at Agate and years of research in the laboratory by numerous paleontologists. When the first excavations were made at Agate there were only preliminary inklings of how unusually rich the fossil deposits there were. Yet these early insights were sufficiently penetrating to excite visions of what might be expected in future years. Osborn, who was mightily impressed by the collections made at Agate by the Carnegie Museum of Pittsburgh, the University of Nebraska, and Amherst College during the years 1904–7 was especially anxious for the American Museum of Natural History to enter the field. During the summer of 1907, in response to an invitation from Captain Cook, he made his first move.

For much of the summer of 1907 Bill Thomson had been working in South Dakota with his brother-in-law, Paul Miller, continuing the field program of the previous summer when Thomson and Will Matthew had sweltered in the high plains rock exposures along Porcupine Creek. This year success was indifferent, and at a moment when both Bill and Paul were feeling particularly discouraged with their results, a letter arrived from Professor Osborn, dated August 13, which directed Thomson to move to Agate, where the Professor hoped that "we shall make a very strong finish" for the collecting season, and thus offset the disappointing results of the work in South Dakota. Bill and his party immediately vacated the Dakota field, and on August

17 Thomson wrote Matthew that the crew was on its way to Agate, Nebraska. "With the party we have I think that is the best plan as we can surely find something there, besides I hope we may be able to open up that quarry for some good work for next spring" (letter of Thomson to Matthew, August 17, 1907). Thomson's preliminary excavation of 1907, on the north side of Carnegie Hill, quite removed from Peterson's "southwest excavation," was the beginning of a long campaign in the northwestern corner of Nebraska—a campaign that would continue for two decades, through all the remaining years that Matthew was associated with the American Museum.

In the same letter Thomson remarked that "Dr. H. [W. J. Holland, director of the Carnegie Museum in Pittsburgh] seems to have queered himself with Mr. Cook." Thereby hangs a tale.

The early work at Agate had been carried forward, as noted, by Professor E. H. Barbour, Professor Frederick B. Loomis, and O. A. Peterson. Peterson was a good man, but he had the misfortune of having to work under the thumb of Dr. Holland, who was a very acquisitive person. At an early stage in the Agate exploration he became obsessed with the idea that the Carnegie Hill was his own private bone-hunting preserve, in spite of the fact that he was there by invitation of Captain Cook.

Captain Cook, a most generous person, wanted the institutions interested in making collections of fossils from the Agate deposits to work there freely and in harmony. But in 1907 he was more than a little annoyed by Holland's grasping tactics and, as mentioned, invited Osborn, representing the American Museum, to come to Agate to partake of the paleontological riches of that remarkable site. Of course, Osborn was delighted to accept the invitation.

Holland was not pleased and, being a strong-minded man, attempted to scuttle Captain Cook's plan for cooperative work at Agate. In a letter dated March 31, 1908, Cook described the situation to Osborn.

My dear Dr. Osborn:

I have run against a snag and I write to let you know that Dr. Holland has written me that if other institutions are to be allowed to work at the bone hills he will quit the field at a great loss. He was anxious to buy or lease the quarry so that he might exclude all others from working there, but I do not wish to either lease or sell to him. I tried very hard last year to have all of the three institutions that have worked on the two bone hills to work together in harmony in the interest of science in this locality as the amount of material and the labor required to secure it is so great. It now seems as though my efforts along the line of harmony have been a failure, and I do not know what is the right thing to do.

The thing that was done was to deny the Carnegie Museum collecting privileges after 1908, and this must have been a bitter blow to Peterson, who was caught between two powerful men—Holland and Osborn. Robert M. Hunt, Jr., in his splendid 1984 monograph, "The Agate Hills," gives a succinct presentation of Peterson's dilemma and disappointment.

> Matthew's righteous indignation, Osborn's patrician certitude in firmly excluding Peterson from the quarry, Holland's general failure to support his man—one wonder's at Peterson's apparent patience and fortitude in the face of the seemingly endless irritations visited upon him by events at Agate over the five years he worked in the quarries. To have made such a discovery when none had recognized its value, to have labored in the field to develop the site, mapping carefully blocks and bones, to have written during evenings and weekends to produce the first descriptions of the fossils, and then to have such conflicts materialize would have tried the best of men. Field paleontology was no picnic."[4]

In spite of the frustrations and disappointments that beset Peterson in the field, he was the person who was most productive in the publication of results from digs at Agate during the five years that culminated in the great field season of 1908. Eight contributions describing Agate fossils were written by him, and published in 1905, 1906, 1907, 1909, 1910, 1911, 1920, and 1923, and a joint publication with Holland appeared in 1914. Four of these publications were large monographs: on the giant piglike entelodonts—1909; on the carnivores—1910; on the rhinoceroses—1920; and (the joint paper) on the chalicotheres—1914. He was not to be kept down by the machinations of his superiors.

---

[4] Hunt, ibid., p. 105. Various aspects of the history of paleontological work at Agate are derived from Dr. Hunt's publication. Dr. Hunt also provided an extensive series of photocopies of notes in the field books of Thomson.

The sad story of the protracted disagreements between Holland and Osborn has been fully recounted by Hunt and need not be repeated here. Suffice it to say that Will Matthew had the final word, in a memo written at the behest of Osborn concerning a long self-justifying letter written to Osborn by Holland on November 9, 1908, in which Holland relinquished what he considered his rights at Carnegie Hill to other institutions, specifically to the American Museum. The final two sentences of Matthew's memorandum read "I have no doubt that Dr. Holland is aware that Mr. Cook has no intention of letting him work any more in that quarry. He does wisely therefore in yielding gracefully what he cannot possibly hold."[5]

In spite of Captain Cook's despair over his inability to have the various people digging at Agate work in harmony, and in spite of the 1908 field season being the last for the Carnegie Museum, that summer was perhaps the most prolific season of institutional work in the history of the Agate quarries. Peterson was there with three assistants, making his last big haul from the quarry where he had in previous years found such rich concentrations of fossils. Will Matthew and Bill Thomson were there working on the opposite side of Carnegie Hill from Peterson, and making their first significant collections for the American Museum. Barbour was there with a crew of four, excavating at University Hill for the University of Nebraska. Richard Swann Lull and his faithful assistant, Hugh Gibb, were there, also excavating at University Hill, as well as at the *Stenomylus* quarry, for Yale. And Loomis with four assistants was also at the *Stenomylus* quarry. A picture taken by Bill Thompson in 1908 shows the tents of the several field parties dotting the prairie north of the Agate Hills, and according to the stories that have come down from those days, they were enjoying one anothers' company and having a good time.

One significant development at Agate in 1908 was Will Matthew's first meeting with Harold Cook, a crucial figure in the history of the Agate Hills quarries; a close and cordial relationship developed among Matthew, Thomson and Cook.

Harold Cook was born on July 31, 1887, in Cheyenne, Wyoming. When he was six weeks old his parents, as mentioned, brought him in a buckboard (a light wagon pulled by two horses) from Cheyenne

---

[5.] Hunt, ibid., p. 107.

to Agate, a journey of about a hundred miles. It must have been an arduous trip for a young mother with a new baby, over rough dirt roads and under the blazing late summer sun. Thus Harold came to what was then known as The 04 Ranch—subsequently to be called the Agate Springs Ranch, and there he lived as a boy and throughout much of his adult life.

He grew up in the spacious environment of the high plains, and spent many of the days of his youth on horseback, with a hunting rifle at hand. But being his father's son, he soon became inordinately interested in the fossils that were being discovered and excavated from the two hills during the first paleontological expeditions to Agate. Indeed, he spent many hours working with the paleontologists, who were bringing to light the masses of bones that had for so many eons been buried in the rocks down river from the ranch house. He thus developed a remarkable familiarity with the fossils, even as a lad, and was inspired with an ambition to lead the life of a bone hunter.

It is not at all surprising, therefore, that when Harold Cook reached the proper age for college, he went to the University of Nebraska in Lincoln, to study geology and paleontology under Barbour and his associates. He was still at the university in June of 1908 when Matthew arrived at Agate, and several days elapsed before he finished his academic year and was able to join Matthew and Bill Thomson, who were waiting for him at the ranch.

Matthew, getting acquainted with the ranch and the people there, and looking over the bone quarry and the surrounding countryside, began to develop several plans for a campaign at Agate—not only at the quarry but also in the hills beyond the ranch, where fossils were to be expected. Another plan was also forming in his mind, a plan that intrigued him, but one which he reluctantly had to abandon.

He outlined this idea in a letter to Kate, written while he and Bill Thomson were waiting for Harold to arrive.

> The plans are very unsettled as yet, partly because Mr. Cook does not want to make any arrangements until Harold comes (he is expected tomorrow) and partly because I want to see the work taken up if at all in a more thorough and efficient way than has been done. My idea is to make a permanent camp here, build a shack and employ two men throughout the year, working the quarry and prospecting in the summer, preparing specimens here

through the winter and sending to the museum such prepared material as was wanted there, keeping the rest here for a sort of local museum subsidiary to the American Museum. This of course would need financing and Professor Osborn might not endorse the plan, or, approving, might not be able to raise the money.

(letter to Kate Matthew, June 7, 1908)

This interesting, innovative idea never came to fruition. Perhaps it was too bold for its time; subsequently, it was made obsolete by technological advances in our culture that Matthew had no way of foreseeing in 1908. First, the automobile rapidly developed, and after that the airplane.

In 1908 the automobile was a pretty primitive affair, and the airplane had barely gotten off the ground. Will Matthew and his fellow paleontologists in those days traveled by train to their collecting areas and then made their way across bumpy terrain with horses and wagons, which was slow business. Of course it made sense to have a preparation laboratory on the spot if a lot of travel could be eliminated, but within a few years—hardly more than a decade—it made better sense to rely on the automobile in the field and send the fossils back to the museum where they could be prepared and studied according to established priorities. Travel to the collecting areas still was by train, but within a few decades the airplane made all fossil sites relatively near at hand—even those on the opposite side of the earth.

Another modern development that Matthew could not have foreseen was the incorporation of many fossil sites into the National Park System, and the establishment of special museums, *in situ* exhibits, and laboratories at those sites. Agate is a case in point. Today the quarry and surrounding region is a national monument, now protected by the government and made available for the instruction and enjoyment of the public.

The day after Will wrote Kate of his idea, June 8, Harold Cook arrived at the ranch and in short order a paleontological campaign at Agate was inaugurated by a new triumvirate—Matthew, Thomson, and Cook. (Granger, it may be recalled, was many miles away collecting Eocene fossils in the Bridger and Bighorn Basins.) These three worked closely together during the next two decades, and the ties between Bill Thomson and Harold Cook lasted for many years after Matthew's death in 1930. In 1908 the association of Will with Harold was just beginning, and the two men were developing a friendship that would be the basis of their work together in northwestern Nebraska and on Nebraska fossils for years to come. To quote again from Will's June 7 letter to Kate:

> Tomorrow we are all invited to dinner at five o'clock with the Cooks—dinner to celebrate Harold's return from college. He is going to work for us during his spare time, and has the reputation of having found most of the material that has been obtained from here. Osborn thinks very highly of him. Cook [James Cook] is an interesting man, a well-to-do ranchman, well-informed, travelled, acquainted with a great many noted men, and in the fossil material, he is not looking to make any money out of it, but wants to have it worked up and described for the benefit of science and his own personal interest. Also Harold figures largely—his father wants to have him get as much interest and benefit as possible out of the material. Altogether I figure that at present the main thing is to get well in touch with the Cooks, to explain our plans, methods and ideals as fully as possible, and to arrange so that we can swing into the permanent camp plan later on if it proves feasible.

Captain Cook was delighted that there should be five institutions all working at Agate in that summer of 1908. On more than one occasion he had expressed his wish that the bone deposits in the two hills might be thoroughly excavated during his lifetime, so that the fossils could be widely distributed among museums.

In 1908 Will Matthew was concerned with more than the existing Agate digs; most of the American Museum work there was to come later. He was exploring various sites in the vicinity of Agate and was planning to make an extended prospecting trip to an area some fifteen miles south of the ranch, where Harold had found some interesting fossils.

In the meantime he wrote Kate, "Osborn has postponed his return from Europe I understand, and so I don't know when I'll hear from him about the permanent camp idea. Meantime I'm going next week to look into the country south of here and if we find material may not be back. I plan to go first to the head of Snake Creek, then to Spoon Butte, then to the breaks north of the Platte, and in one of those places hope to find some good prospecting ground. This country is getting a little too crowded for me" (letter of June 25, 1908).

Here one sees the germ of a project that was to loom large in Matthew's paleontological career, namely, that of Snake Creek. The best way to understand and appreciate the beginning of Matthew's paleontological work at Snake Creek—the name was used, in an inclusive sense, to designate the accumulations of small rock exposures in the high prairies to the south of Agate—is to reproduce two letters to Kate.

> Agate, Neb.
> June 29, 1908

Harold Cook and I start this morning for a short trip, down to the exposures 'round the head of Snake Creek south and southeast of here about fifteen miles. So you won't hear from me for a couple of days after this when I hope to be back again. We are going to reconnoitre things chiefly, and not to collect, but I want to have country to work in through the latter part of the season and next year, and this is to look for it.

> Agate, Neb.
> July 3, 1908

Here I am back after a four day trip to the southward with Harold Cook. We got some surprising and quite unexpected results, and made a big haul of fossils, all fragmentary but very interesting. We located two new fossils deposits, one of which contains a fauna of some thirty species and is apparently of the Pliocene age. The Pliocene fauna is very little known in this country, and what we obtained will double the number of animals and amount of material that has been found hitherto in the U.S. The other fauna is of Middle Miocene age, and is new to this part of the country, and promises to yield some good specimens. The Pliocene fauna has some very interesting species in it, about half of them are probably new, and some are of huge size. An enormous camel, as big as a giraffe, is very common, and a huge carnivore bigger than a lion is represented by some two or three bones. So you see it is of a good deal of scientific interest. We brought back a great quantity of separate teeth, jaws, foot-bones, incomplete limb bones, etc., and are going down with the whole outfit in a week or so to round up all we can get out of it.

To return to our trip; we started out Monday morning with the light rig, and went down about twenty miles to the south till we came to the broken country facing the Platte River. It is all rolling plains between, with occasional sandhills some a couple of hundred feet high all made of up drifted sand, but mostly grassed over, and all the plain covered with bunch grass. This grass grows only a couple of inches high in little tufts and intermingled with it are grasses of larger size, but very little of the grass over six inches in height. Here and there are windmills for pumping water from the driven wells, they can get water almost anywhere on the plains at depths of from fifty to five hundred feet. The stock water at these windmills, and we watered our team and filled our canteens at one or another of them.

Our walk was along the top of the "breaks" (country cut up by gullies and draws opening down into the Platte Valley.) The beds exposed here were thick with fossils toward the top but very barren in the lower strata. The second night out came a heavy drenching rain which we received on our "tarps"—we had a small shelter tent but did not use it except on the first night which was so cold that we would have been nearly frozen outside. The rain didn't hurt any, although it rained off and on all the next day and the next two days were bright and clear and very hot in the middle of the day. We lighted our fire with a little alcohol poured on a "cow-chip" (this is the only kind of fuel you can get on the plains) and got along very well, as we had taken the precaution to put a few "chips" under cover from the rain. It was very pleasant to get to the real thing in camping out for a change, and Harold Cook is a fine fellow to be with. He is a remarkably clever young fellow just under twenty-one, an enthusiast in paleontology and well up in all the plainsman's accomplishments. It was delightful to see his enthusiasm over these new faunas and the quick way in which he grasped the significance of each specimen as we picked it up. He has a good instinct for the stratigraphy too and readily worked out the relations of the different beds we were working in. And a fellow of fine ideals and good culture. I hope we'll get him to New York this winter, and we'll see as much as we can of him if he comes. He would fit in fine with our bunch.

Tomorrow Mrs. Cook has invited all the scientific parties to meet at her place, and there will be tennis, and some wrestling between Harold, Stein and one of the University of Nebraska men, while we who have no accomplishments will sit 'round and talk and take it easy.

Stein, a Finn, was the camp cook, and very much appreciated by Will Matthew and his crew, because in Matthew's words he was "rather a crank on wholesome foods," a welcome change from some of the cooks that had been endured on previous expeditions.

The Fourth of July picnic was a jolly affair, with the representatives of three museums on hand: the American Museum, the Carnegie Museum, and the University of Nebraska Museum.

> 15 miles south of Agate, Neb.
> July 13, 1908

> We started out Saturday morning, rather late, and found it necessary to camp early, as there was a downpour of rain getting ready. Had things fixed in good time and managed very comfortably. Yesterday we started on again and made our camp here about noon. In the afternoon came another downpour and did not let up until five o'clock, when we went over to the place where we have planned to quarry, and started in. Worked all day today on it, with fair results—about what I had expected. I think we'll stay with this a few days longer and then do some prospecting around while two or three of us keep up the quarry work. With the whole outfit working in the quarry I think we can average 500 horse teeth and two or three complete jaws or limb bones a day—which is fair results. By the end of the season, working in these beds, we should achieve quite a collection—at all events a good many thousand horse teeth, and as there are at least eight or ten species of fossil horse among them, we need a very large series in order to work out the characters satisfactorily. They are all three-toed horses, about as large as a sheep and up to the size of a cow.

Matthew's remark to Kate about collecting several hundred horse teeth each day, as well as jaws and limb bones, is a clue to the nature of the fossil deposits at Snake Creek. These fossils had been accumulated in river channel beds, and as such are abundant but fragmentary, having been rolled about and broken by the action of the water.

When a paleontologist comes along, he can therefore gather fossils by the basketful. This may not sound helpful to someone searching for complete skeletons, but to fulfill the needs of certain paleontological studies, such as the analyses of populations or species, having at hand large numbers of, for example, horse teeth, is remarkably valuable. Especially when one considers that much can be learned about the evolution of horses from the detailed study of their teeth.

Of course not all Snake Creek fossils were so fragmentary. Many skulls, jaws, and even skeletons were found, especially in the flood plain deposits through which had been cut the channels where the horse teeth accumulated. Taken together, all the fossils, complete and incomplete, furnished an excellent record of a fauna, giving a reasonably balanced view of what mammals were living together in the interior of the continent.

In mid-July Will wrote to Kate, assessing the results of their early work and describing some of the problems of collecting and of camp life at Snake Creek.

> In camp 20 miles south of Agate,
> July 16, 1908

Harold is going to Agate this morning, so I'll take the chance and send you a line by him.

We worked hard the last three days and as a result have gotten together a good many jaws and thousands of teeth, and increased the list of species to about thirty-six. One skull and jaw of a big camel (but not the largest size) with limb bones and vertebrae associated.

The cattle have been a good deal of a nuisance. They come around camp during our absence and start investigating. We have had to sew up our tent in half a dozen places where they tore it, and they have messed up things a good deal. Otherwise we are pretty comfortable. I think we'll be here for some time, as the beds are turning up pretty good material. This will be certainly a valuable collection scientifically tho' I don't see much exhibition value in it yet.

> In camp 20 miles S. of Agate, Neb.
> July 18, 1908

I am sending in this letter by Stein to Agate, as he is going in with a load of fossils, to bring out provisions. We haven't made any special finds since I wrote last, but have been working on the pockets of fragments with fair success. This collection will yield a lot of new species, and a great quantity of horse specimens for exchange and sale, so that we ought to get some interesting results out of it. All of us will turn gophers if we keep at it long, as we claw away at the sand with pick and awl and scratch up the loosened stuff with our hands. I expect to have well developed digging claws on my fingers by the time I get home. Wish you could see the quantity of petrified teeth of three-toed horses that we have got together. There are several thousand of them, representing probably a dozen species. We have a large series of jaws of various animals—some forty species altogether, and quantities of limb and foot bones. There are very pretty specimens of petrified wood in with the bones.

Agate, Neb.

July 25, 1908

I'm really on the way at last, and have got so far as Agate at least. This afternoon I take the mail-wagon to Andrews, and train from there east, as per schedule in my last letter.

When Will Matthew boarded the mail wagon to Andrews on July 25, he had several things to think about. The plans for future work at Agate and Snake Creek must have been predominant among his thoughts, but he very likely was thinking as well about Harold Cook. He had spent part of the summer with Harold and found him to be a young man much to his liking. Harold was a very lively person, dedicated to the search for fossils, but interested as well in the general natural history of the high plains where he had grown up. He had a pleasing personality, he was cheerful, and he had a rollicking sense of humor. Moreover, he was physically strong, as might be expected of an outdoors person, and he was a hard worker. It is significant that as early as July 3, in Will's letter to Kate, he had stated that he hoped to "get him to New York this winter."

In his autobiography, *Tales of the 04 Ranch,* Harold Cook wrote of Professor Osborn's visit to Agate.

> He told us that evening as we sat on the porch, that the following year would be his last before he retired from active teaching. . . .
>
> "I want Harold to come back to Columbia and take my class, also the classes of W. K. Gregory, the comparative anatomist," he said.
>
> Dr. Gregory had been at Agate with the American Museum party: all of us knew him and admired him. The idea of attending Columbia University thrilled me. . . .
>
> Dr. Osborn offered me the job of identifying the Snake Creek collection with Dr. Matthew. If nothing else could have won me,

that offer would have.... Dr. Osborn said ... he would be my sponsor, personally, and would be my special adviser in graduate school.[6]

Captain Cook was reluctant about Harold going so far away from the ranch, but Harold's mother championed the idea, with the result that Harold did spend that winter in New York, with Osborn, Matthew, Granger, and Gregory. The bonds of his friendship with Will Matthew were strengthened.

"In the half day [each weekday] I spent at the American Museum," wrote Harold, "I was collaborating with W. D. Matthew, one of the world's finest paleontologists, who had an encyclopedic knowledge and memory. He knew every mammal, living and extinct, that had ever been described, including the characteristics by which it had been differentiated, and who described it. All this he knew offhand, as well as I knew my own name."[7]

For Harold the academic year spent in New York was a time of intellectual adventure and of new and challenging experiences for an unsophisticated and rather raw young man, fresh from the sparsely populated hills of northwestern Nebraska. And because of Professor Osborn's sponsorship he was on occasion thrown in with a class of folk who must have seemed almost foreign to him. For example, he told of a social affair at the Osborn home, Castle Rock, on the Hudson River.

On Halloween there was a party at the Osborns, with forty or fifty guests, the Theodore Roosevelts among them. I rented a dinner coat and went, for I wanted to meet Teddy Roosevelt. Alice Roosevelt and Josephine Osborn were close friends; at about midnight they all decided to go down to the kitchen for an impromptu meal. I was helping prepare scrambled eggs when Alice came up behind me, reached around, and plastered a handful of white flour in my face.... I had heard about the roughhousing carried on by the young Roosevelts, and I had very good reason to remember "Princess Alice."[8]

[6] Cook, *Tales of the 04 Ranch,* pp. 200–1.
[7] Ibid., p. 203.
[8] Ibid., p. 204.

At the end of the academic year Harold returned to Agate with a greatly expanded knowledge of fossil mammals and a reinforced feeling of loyalty and respect for Dr. Matthew.

As for Matthew and for the museum, there was now an enthusiastic ally, ready to support with all of his ability the museum's program at Agate. This was important, because by now Harold was, in the eyes of Captain Cook, a full-fledged member of the paleontological profession and his advice was to be heeded by his father. Consequently, a strong association was developed between the Cook family and the American Museum, as represented by Osborn, Matthew, and Thomson, which led to significant advances in the development of paleontological exploration in the high plains.

One immediate result was the inauguration of an extended program of excavation at Carnegie Hill, and subsequently at Snake Creek, under the direction of Albert Thomson. He began his new quarrying operations at Carnegie Hill in 1911, carried them through 1914, began again in 1916 and worked through 1920, and quarried once more at Agate in 1923. These activities were interspersed with work at Snake Creek in 1916, 1918, 1921–22, and—after 1923, his last year in the Agate quarry—from 1924 until 1927.

Bill Thomson's years of quarrying at Agate yielded two paleontological collections of unsurpassed value, the first scientifically significant, the second of supreme exhibitory value. The first collection consisted of seventeen skeletons of *Moropus,* the horselike chalicothere with clawed toes instead of hooves, collected at Carnegie Hill between 1912 and 1916. Chalicotheres are not common in the fossil record; consequently, a small population sample of these ancient and curious mammals afforded the opportunity to study individual differences resulting from variations in size, age, or sex. It was a sample of chalicotheres the like of which had never been seen.

This sample of a chalicothere population intrigued Osborn, who made plans to study the fossils and publish a monograph describing them—one consistent with other studies of perissodactyls that Osborn had done or had in mind. But the description of Agate chalicotheres at the museum never eventuated, which is a pity. It seems likely that the press of other projects in Osborn's busy life crowded it out. As it turned out, the large, definitive study of the chalicotheres from Agate was published by Holland and Peterson in 1914 as a Memoir of the

41. Professor Osborn in 1921, at age sixty-four.

42. Osborn (*left*) and Matthew (*right*) in the preparation laboratory of the American Museum in 1923, inspecting the skull of *Baluchitherium,* a gigantic rhinoceros found in the Miocene sediments of Mongolia. Otto Falkenbach (*seated*) and Carl Sorensen, paleontological technicians, are holding pieces of the skull in their proper positions. Osborn is characteristically clad in a long, formal coat.

43. William King Gregory, the gentle scholar and world authority on the vertebrates or back-boned animals, about 1920. When this picture was taken he had succeeded Osborn as professor of vertebrate evolution at Columbia University and was actively engaged in the study of primate evolution. He strongly opposed some of Osborn's views concerning primate history, especially with regard to the evolutionary lines leading to *Homo sapiens,* yet he and his former professor never allowed their divergent views to cloud their close friendship.

44. Henry Fairfield Osborn, the well-dressed explorer, in 1919. Osborn was never an active field man—he quite frankly admitted this— but he liked to visit collectors in the field.

46. Dr. Matthew at his desk in the American Museum of Natural History, evidently at an early stage in his paleontological career. As may be seen, there was nothing ostentatious about Matthew at work. From his cluttered desk, jammed between storage cases, he could swing around 180 degrees to his equally cluttered work table.

45. An early field party of the American Museum of Natural History in 1895, when two members of the triumvirate were very young. *Left to right:* O. A. Peterson, Jacob Wortman, Walter Granger, and Albert Thomson. On this expedition Wortman and Peterson, aided by Thomson, were collecting fossil vertebrates. "Bill" Thomson was also camp cook, photographer, horse handler, and general handyman. Walter Granger was collecting modern birds and mammals—this being before he became permanently connected with the Department of Vertebrate Paleontology.

47. Walter Granger in 1911, by now an accomplished fossil collector and student of ancient mammals.

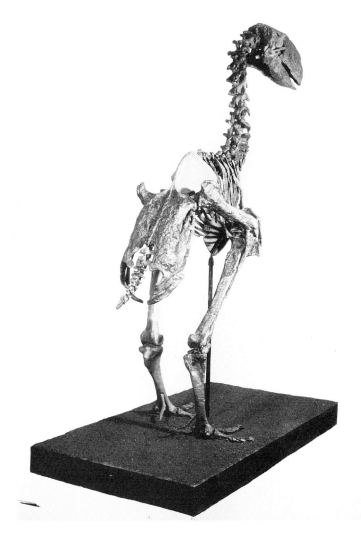

48. One of the great discoveries made by Walter Granger was the skeleton of a gigantic flightless bird, *Diatryma,* found in the Eocene Wasatch beds of Wyoming. It was described by Matthew and Granger in 1917. They showed that this huge bird, more than six feet tall, was a dominant predator (as indicated by its extraordinarily large beak) that occupied an ecological role only later taken over by the large carnivorous mammals.

49. Walter Granger in the field at the time he was collecting Eocene mammals in Wyoming, which would form the bases for the epochal series of studies by Matthew and Granger on the Wasatch and Wind River faunas. Those were the days when every self-respecting fossil collector working in western North America was necessarily a horseman.

50. The home of Captain James Cook at Agate, Nebraska. This comfortable house, set within an extensive grove of trees lovingly planted and nurtured by Captain Cook, was for decades a haven for fossil collectors working at the Agate Hills. Today Agate is a national monument.

51. Captain James Cook and Albert Thomson, two remarkable men, at the Agate fossil quarry. Captain Cook, a redoubtable scout, plainsman, and rancher who spoke the Dakota language fluently, had a most enlightened policy concerning the fossils found on his land. He made the Agate Hills available to all accredited museums, so long as their paleontologists behaved like gentlemen. "Bill" Thomson, one of the triumvirate, spent many summers at Agate, where he made superb collections and produced invaluable field records. He was an accomplished photographer: note the shadow of the camera on a tripod, which indicates that this was a self-timed picture of the two men.

---

53. Fossil hunters' camp at "East Agate," at the foot of the Agate Hills, some three miles down the Niobrara River from the Cook house. Harold Cook homesteaded this land and built the cabin, which was made available to fossil hunters through the years. When I camped here with my crew in 1938, we parked our field cars by the cabin, where formerly only horses and wagons stood. This is now part of Agate National Monument.

52. The Agate Hills. University Hill is on the left, Carnegie Hill on the right. A broad, light-colored streak halfway up Carnegie Hill (above the team and wagon) marks the site of quarrying operations. A similar quarry site is seen on University Hill.

54. University Hill, seen from Carnegie Hill. Note the quarry platform on University Hill and the paleontologist (probably Bill Thomson) at the lower right edge of the picture, looking up from the Carnegie Hill quarry. A primitive Ford is parked between the hills.

56. Working in the Snake Creek quarries, south of Agate. These were small "digs" spotted around the landscape. Mrs. Thomson was there, helping Bill. Just below the horizon is the Snake Creek camp.

55. Harold Cook, son of Captain James Cook. Will Matthew was extraordinarily fond of Harold and was almost a surrogate father to the young man, whose abilities he respected whole-heartedly.

57. William Diller Matthew (1922). Portrait by Lewis
Hine. Lewis Hine is renowned in the annals of American
photography as a depicter of social and industrial scenes.
The Hine and Matthew families were friends and neighbors in Hastings-on-Hudson. Lewis Hine, departing
from his usual subjects, photographed the members of
the Matthew family.

58. The completed south facade of the American Museum of Natural History—the combination of buildings in which Matthew worked during his years in New York. His office was on the top full floor of the square building at the right end of the complex. Osborn's offices were on the same floor and included the round tower to the right, with the balcony around part of its periphery. (Today the museum is a vast complex consisting of twenty-two sections, each a sizable building in its own right.)

59. A California picnic. *Left to right:* Professor and Mrs. Bruce Clark, Charles and Jessie Camp and their son Charles, and Kate and Will Matthew.

60. Dr. Matthew with graduate students on a field trip in Nevada, May 1928. *Left to right:* Sumner Evans, R. A. Stirton, Matthew, V. L. VanderHoof. "Stirt" and "Van" were Matthew's prize graduate students. After Matthew's death Stirt succeeded him and for many years taught mammalian paleontology and researched fossil mammals. Van, who was particularly close to the Matthew family, held several positions and eventually became director of the Santa Barbara Museum.

61. A bust of William Diller Matthew done by his daughter Margaret Matthew Colbert. Several original casts of this sculpture were made; one is at the American Museum of Natural History, one at the University of California at Berkeley, and one at the Museum of Northern Arizona, Flagstaff.

Carnegie Museum, entitled *The Osteology of the Chalicotheroidea*—a large format work of 217 pages, including numerous fine plates. For the American Museum an opportunity for publication of a significant study of Agate fossil mammals had been missed.

During his excavation of 1913 Thomson found a skull of *Dinohyus,* the gigantic piglike entelodont, a tantalizing discovery for all concerned, because Matthew and Osborn had been hoping from year to year that a skeleton of this remarkable beast might come to light, to enhance the museum collections. But no skeleton was found; even today only two skeletons exist—those collected by the University of Nebraska and the Carnegie museum.

The other great find that Thomson made at Carnegie Hill was a superb concentration of rhinoceros bones that he excavated in 1919 in the form of a block or slab, eight and one-half feet in length, four and one-half feet in width, which weighed more than two tons. This slab, now at the American Museum in New York, is probably the most spectacular Agate block to be seen.[9]

It is curious that, with all the fine work carried out by Bill Thomson, supplemented by his careful notes describing the details of deposition at the quarry, so little follow-up in the publications of the American Museum took place. We have seen that Osborn failed to proceed with the monograph on the chalicotheres. And Matthew's interests were focused elsewhere—specifically on the fossils that he and Thomson collected at Snake Creek. Thus the most visible and permanent result of the decade of American Museum excavations in the Agate quarries is Thomson's rhinoceros bone slab.

Various hypotheses have been put forward to account for the remarkable concentration of bones in the Agate quarries. Peterson, in 1909, postulated that the bones had accumulated at a river ford. Holland, in the chalicothere monograph of 1914, attributed the concentration of fossils to the gradual accumulation of bones at a pool on a floodplain, where animals came to drink and there died of various causes, including attacks by predators.

In 1923 Matthew presented his quicksand hypothesis in a short, semipopular article in *Natural History,* which, as Robert Hunt has

---

[9] The slab has been on exhibit there for years. However, as of this writing, the museum is remodeling its fossil halls, and they are presently closed to the general public.

pointed out, "emerges as the principal account of the Agate work by an American Museum paleontologist."[10]

The formation is a rather soft sandstone of light gray color, made by the accumulated floodplain sediments of a river that flowed eastward across the plains, for then as now the region was one of open country and grassy savannas. It is believed that the accumulation of bones was formed in an eddy in the old river channel at a time when the valley was not so deeply cut out as it is now and when the river flowed at the higher level. A pool would be formed at this eddy, with quicksands at its bottom, and many of the animals that came to drink at the pool in dry seasons would be trapped and buried by the quicksand. The covering of sand would protect the bones from decay and prevent them from being rolled or waterworn by the current, or from being crushed and broken up by the trampling of animals that came there to drink. But sand of this kind is always moving and shifting (whence its name of quick) and with it the buried bones would be shifted around, disarticulated, and displaced, so that when finally buried deeper by later sediments of the river valley, they would be preserved as they are found here, complete and almost undamaged, yet all separate and dissociated.[11]

Since 1923 other theories have been put forward, including Hunt's proposal that the bones at Agate are the result of an "unknown death event" which led to the accumulation of many carcasses in ephemeral streams.

10. Hunt, "The Agate Hills," p. 41.
11. W. D. Matthew, "Fossil Bones in the Rock," *Natural History,* 23 (1923):359.

CHAPTER 16

Snake Creek by Car

*M*atthew's next visit to Agate and Snake Creek after the 1908 season was in 1916. In this year Matthew was introduced to the use of the automobile in the field. In a letter from Valentine, Nebraska, dated July 14, 1916, he told Kate "I know you'll enjoy running 'round in the auto, a lot more than I do, for there are a good many other things that I'd rather be doing. But it certainly does get you about from place to place. We've covered ground during this summer that would have taken two or three years to get over with a team and outfit." This seems a rather cautious but favorable acceptance of the modern revolution in transportation.

Ten years later Matthew was fully converted.

> The coming of the automobile has revolutionized the fossil-hunting business. For one thing, it has greatly widened the range of practical field work. The old conditions limited it to a radius of five or ten miles from water and feed. A 'dry camp' supplied at intervals with food and water might carry the exploration a stage further, but throughout the West, especially between the Rockies and the Sierras, there were enormous areas of bad-land exposures that it was not practicable to prospect adequately for lack of water or feed. With the automobile there is probably no promising ex-

posure so distant but that it can be, and will be prospected for
fossils. Moreover, the range of a day's collecting with an automobile
is so greatly increased, that most areas in the Plains can be prospected from some near-by settlement, and it hardly pays even to
camp out.[1]

It may seem that much—perhaps too much—emphasis is being
placed on the use of these early automobiles in paleontological fieldwork. Yet this is not a trivial matter in the history of paleontology. To
a late Victorian-Edwardian man like Matthew the automobile in the
field was a marvelous adjunct to his exploring and collecting efforts.
The opportunities for discoveries were expanded enormously, and for
this Will was most appreciative. From 1916 on Will Matthew utilized
a car for his field explorations, but it had to be a car *with* a driver. His
eyesight was so poor that he never attempted to learn to drive.

George Gaylord Simpson, Matthew's successor at the American
Museum of Natural History, told an amusing story in this connection.
A young man who had just completed his graduate studies, he was
scheduled to be Matthew's field assistant in Texas in 1924. He met
Matthew in Texas, where they waited for the delivery of a new Model
T Ford that was to be their field car for the summer. The car ap-
peared, Matthew and Simpson got aboard for a test drive outside the
little town where they were staying, and off they went, after a fashion.
There was one problem: Simpson did not know how to drive.

Fortunately, they were on a lonely, dirt road out in the country, and
George Simpson improvised as best he could. He never said a word
to Matthew about his inexperience as a chauffeur, and Matthew never
realized in what peril he had been. Simpson quickly mastered the
simple procedures involved in piloting a Model T across the prairie,
and so the summer went well after all.

Earlier, in 1922, Matthew was back at Agate and Snake Creek,
with Bill Thomson as his right-hand man. It was to be Will's last
summer in northwestern Nebraska, whereas Thomson, as noted, continued the work there in later years. The main objective in 1922 was
to continue the Snake Creek work, which was done with much success, but there was a secondary goal—to try to gain access to fossil

[1] W. D. Matthew, "Early Days of Fossil Hunting in the High Plains," *Natural History*, 26
(1926):454.

sites on the Ashbrook land in the Snake Creek area. In contrast to the other ranchers of the region, Mr. Ashbrook was adamant about not letting fossil hunters on his property—just across the fence line from where they were working. He had no particular reasons for this, except that he didn't want people on his land. And Matthew was there to try to bring him around.

Will's first letter to Kate, from Agate, mentioned this.

> Agate, [Neb.]
> July 5, 1922

Reached Agate on the evening of the third, and yesterday, the fourth of July, there was a big neighborhood celebration at the upper meadow a mile above the house. About three thousand people came, mostly in automobiles. They had a barbecue; a big steer was roasted whole in a great covered pan, like a dutch oven. Then in the afternoon they had bucking contests, with a prize, not for the men who stayed on the horses, but for the worst bucking horse. Dancing in the evening on a big wooden platform with a pavilion set up over it and gasoline flares and a very weird band. Various other attractions. We went up mainly in order to see Mr. Ashbrook, the owner of the land where Thomson wants to work, who is still recalcitrant. It will take time and careful handling to get this matter arranged. Meantime we are working on other parts of the exposures, not so promising, but which we have the permission of the owners to work. I have been staying for a day or two with the Cooks, who are as cordial and hospitable as ever. They have four children now, all girls, the oldest perhaps halfway between Margaret and Elizabeth, and the youngest about four. Captain Cook, Mrs. Graham (Harold's grandmother) are staying with them, and they have a young girl of about Elizabeth's age also staying, partly as friend of Harold's wife, partly as helper for the children (so far as I can make out). They haven't any servants now at all, and I very much suspect that they are pretty hard up, as things have been very hard on the farmers and ranchmen lately. Thomson is staying down at the quarry shack about four miles below here. His wife and sister are with him, and young Jerry Black. I'll probably join them tonight or tomorrow.

He did move to the cabin, built by Harold Cook some years previously at the foot of Carnegie Hill and affectionately known to the Cook family as "East Agate." From there he and Bill Thomson and an assistant commuted by car to the Snake Creek exposures. It is evident that by now Matthew was a confirmed internal-combustion-engine paleontologist. No more horses and wagons for him!

Agate, Neb.
July 11, 1922

Our present routine is to get up at 5 a.m., have breakfast and get things in shape, then start over for the Snake Creek exposures which are about 25 miles from here. It takes an hour and a half to get there; so we are working it like commuters. We are in the shack near the big Agate quarry, about four miles east of the Cook ranch house. At least the Thomsons are in the shack; I have a tent outside, and the junior member of the party, Jerry Black, has another. His is a little 'pup' tent, and mine a large wall tent. We eat in the shack. When we get over to the Snake Creek exposures we work on little excavations along the line where the fossils chiefly occur. The shovel is the principal weapon of attack as it is a soft half-consolidated sand, which in many places will not keep a vertical wall. The bone is mostly water-worn fragments; teeth are fairly common but complete jaws or skulls are rare.

Some letters tell Kate about collecting in the Snake Creek beds, and about camp life on the high plains:

July 21, 1922
Agate

"Well we haven't been doing anything sensational here, just pegging along on these Snake Creek beds, mostly hard and disappointing work. I get out of it more than the others because my object here was to study the beds, their mode of formation, and the succession of the different strata. That, I am getting some idea about, at all events. We are still held off from the exposures on the Ashbrook ranch, and may not be able to get into those until September (no, you needn't worry *I'm* not going to be here in September). Yesterday we worked in the underlying formation in a little draw where I located a promising pocket. Don't know yet whether it will turn out much as it takes time to prospect it, but it has a showing of a "little bitsy" horse about as big as a sheep; a kind that we have had teeth and jaw fragments of, but no skull or skeleton. Let you know later how it turns out.

It has been hot ever since my bed roll really came (you know the first time I wrote you it had come the postman misunderstood what I was asking about and thought I meant the Michiganders' roll) and I haven't needed but one thickness of blanket over me, if any. In the daytime it is hard on the eyes, and harder on the complexion, hunting fossils in the bare white badland slopes. And my nice whitey-pink complexion that you used to be so proud of, you ought to see what the sun has done to it now. Poor thing! I can see you melt into tears of sympathy, even as I have melted into tears of s . . . , no perspiration I mean. I was strongly tempted to let my beard grow and refrain from haircuts as a defense from the solar sizzlers, but reflected what a weird spectacle I would have become by the time the Fricks came out, and so compromised on a mustache which now shades my upper lip in budding youthfulness so that everyone takes me for my own grandson and asks me "where's your Dad?"

Agate, Neb.
July 27, 1922

As for my health, it's all right enough, but I'm suffering for lack of exercise. Now and then I get some heavy shovelling, but most of the time I'm scratching around with pick or crooked awl and whisk-broom. I need more walking instead of three or four hours a day automobile riding. However, I have made the best of it. I can't get in time for the exercise, but I am doing what I came out to do, getting a definite and clear understanding of the sequence and mode of origin of these formations. It is really quite complex, and the solution of the problem turns upon certain geologic agencies that are not at all understood by most geologists, so that my explanations and theories will, as usual, be regarded as fanciful and far-fetched by most of my confreres. However, they are correct, so I am not greatly concerned over whether people believe them or not. The critical factor is the prairie sod which acts in several ways. First it provides a tough coating and prevents erosion in that way. Second, the underground waters come up to the surface and evaporate leaving a certain amount of lime behind that had been dissolved in them; this lime concentrates therefore in the top layers, about a foot or so down, and serves as a cement to harden the loose sands that have been blown up into dunes by the wind. Third, the grass of the prairie surface catches the dust from the atmosphere and bit by bit is built up into a level plain of "loess." In the meantime the streams bring down mud and sand from the mountains and spread it widely over their flat valleys, the sand and gravel in the channels, the mud away from channels in the still overflow of each flood. The wind in the dry season blows up dust and sand from the channels and mudbanks and piles up long rows of sand-hills on the lee side of the valley, the finer dust being carried further and deposited in the grass of the tableland beyond. The sandhills are pure sand because the wind sorts out and carries off all the finer material, and they lie in heaped up waves like the sea in a storm. Mostly they are covered with grass now, but whenever there is a break in the cover the wind digs out great hollows and piles the material elsewhere.

That's over, but I'll just have time to finish off before we leave. Wind blew hard last night and black clouds to the south when I got up this morning, so I shouldn't be surprised if we ran into some badly washed-out roads between here and the Snake Creek beds. Fortunately, the sandhills do not wash out easily. . . .

Jerry's latest pet is a rattlesnake which he keeps in a barrel outside the door of the shack.

After three weeks of commuting daily between Agate and Snake Creek, Matthew and Thomson finally decided to move south and camp at Snake Creek. Perhaps the four hours of driving every day, on top of eight hours or more of digging in the hot sun, was wearing them down. Certainly negotiating dirt roads, gullies, loose sand, and mud in a Model T Ford must have added a generous ration of stress and strain to the daily life of the bone hunters.

In camp, Snake Creek Beds
August 4, 1922

We moved camp down here last Monday and this is Thursday. We are camped down at the end of a little draw that we call Stonehouse draw, and trying to get a collection out of the Sheep Creek beds. This is an older geological level than the Snake Creek beds, and contains a somewhat more primitive fauna. The commonest thing is a little bit of a horse, about as big as a sheep, but longer legs. We have a lot of jaws and bones of this and two incomplete skulls. Then there is a rhinoceros, two or three small camels, a deer, a pigmy deer, a couple of dogs, one as big as a kit fox, one as big as a red fox, one as big as a wolf. It will be an interesting collection although not much for looks.

We camped out very comfortably on a grassy flat, with three tents, and an awning stretched between the two automobiles. One tent is for the Thomsons, one for Jerry and me and the big tent is the cook tent now. No mail, but I expect we'll send in to a post-office about the end of the week and get our mail forwarded from Agate. Weather has been redhot in the daytime, and with threat-enings of thunderstorms toward night, but none has hit us yet. We are on the Kilpatrick ranch; Ashbrook is not yet ready to let us camp or collect on his ranch, which is the locality we especially want to work.

We live pretty well for camp life—eggs from a farmstead about ten miles away, cost 20¢ a dozen; butter sometimes, potatoes off and on; but the mainstay of life is doughgoods, bacon and ham, rice and raisins, dried peaches and prunes, canned stuff in lesser quan-tity as it is heavy bulky stuff to carry and generally of inferior qual-ity.

This was as far as I had time for yesterday morning. It was furiously hot through the day until about four o'clock and I col-lected a sunburn, working on those whitish rocks, that hasn't quite left me yet. It is annoying that I can't seem to get sunproof by any amount and length of exposure. About 4 p.m. the thunderstorm hit us and we sprinted for camp and had to quit for the day. This morning is bright and cooler with the grass all sparkling and wet, and the birds darting about on the tent and an ambitious steer over the hill pretending he's a bull and bawling with all his might.

> In camp, 20 m.s. of Agate
> Aug. 11 [1922]

We do pretty well for meals here, with butter and eggs purchas-able at nearby ranches; and Thomson is quite a flap-jack artist. Breakfast consists usually of either doughgoods or flapjacks with "surup," bacon, rice (not the kind with the flavor all washed out in the sink) and coffee (boiled, very much so). One can have canned milk and sugar with the rice, and sometimes stewed prunes or apricots—and—peaches. Dinner is about the same and supper similar, except that the biscuit or pancakes are cold and I take cocoa (lukewarm to cold) instead of coffee. It is pretty good as camp food goes, but I'll appreciate some of your getting up just the same, about three weeks from now. It agrees with me well enough, though I don't think it would go as a permanent diet very well unless I took more active exercise than I'm getting at present. But the truth is that you being here would make all the difference. I wouldn't be bothered about food or baths or sunburn or anything else, if you were. But how about you? Could you stand a gallon of water once a week for bathing purposes, and all the various inconveniences of camp life for a whole summer?

In those fiscally innocent days of three-quarters of a century ago, before anything like the National Science Foundation had even been dreamed of, many institutions necessarily relied on self-help to keep their programs alive. The American Museum from the very beginning depended on help from New York City's financial tycoons, and since New York was the great financial center of the nation—perhaps even of the world—there were enough interested tycoons on hand to give the museum much of the help it needed. Osborn was, of course, a member of the so-called New York elite, and was able through his personal friendships to enlist other people in his circle to support the museum programs.

One such person was Childs Frick, the son of Henry Clay Frick, the associate of Andrew Carnegie. Childs Frick was interested in fossil vertebrates, and Osborn encouraged him to channel that interest into the museum. Matthew, in turn, was anxious for Frick to support his program at Agate and Snake Creek, and fortunately Frick was coming out to western Nebraska that summer of 1922 to see what was being done. So in the hot August days, while Will Matthew was scratching and digging in the Snake Creek exposures with Bill Thomson, he was at the same time thinking about the logistics of the Frick visit. He shared his thoughts with Kate:

> August 15, 1922
> Agate, Neb.

> We are planning to take the Fricks up into the badlands north of Harrison on Friday and Saturday and spend Sunday at Agate with the Cooks, then a few days looking over the country round about here, and then strike south and see the Snake Creek country, then south to Scotts Bluff and look into the big exposures of badlands there; then south to Pawnee Buttes where I worked in 1898 and 1901 and then down to Denver, where I'll take the train east to St. John. Will try to telegraph exact date of arrival when I reach Denver and can get my tickets through. It may be a couple of days later than Sept. 1st, but hardly earlier because I don't see how we can get through our program in less time. It's a shame to lose out on vacation this way, but I don't see how we can help it this year. I

simply must get this matter of plans for campaign in this country properly presented to Mr. Frick, because if he takes it up and backs it, it will be a model piece of research of the kind he wants to make, and I have the right men to work with him so as to make it a great success. Professor Osborn, I anticipate, is going to be absorbed in Asiatic research, and will have little time to this problem, much as he is interested in it. And if we are going to make adequate use of the great collections that we have already secured from this region, it will need Mr. Frick's personal work and financial backing to do so. I am disposed to think that if we can arrange it for you to come out here next summer it would be a good act, if we can arrange it for purchase now of the Snake Creek quarry and open up vigorous work there through the summer. However, we'll see about that later. There'd be some fun about bone-digging if you were along.

Mr. and Mrs. Frick arrived and for two weeks were escorted by Matthew and Harold Cook around northwestern Nebraska, where they visited various fossil localities. Mr. Frick thoroughly enjoyed himself and was especially happy one day to find the jaw of a fossil bear, as he and Matthew and Harold Cook were poking around in one of the Miocene exposures. Matthew even hoped to arrange for Mr. Frick to talk with "the recalcitrant Mr. Ashbrook." Whether that happened is not known.

For Will the visit was encouraging, and it inspired him to day-dream about the future program at Agate and Snake Creek. Fate would have it otherwise, as shall be seen; Matthew's plans for collecting and research in north-western Nebraska along the lines he had envisaged never materialized. However, he could dream, and as he rode east on the train to join his family, he had visions of the future. In those visions Harold Cook was a central figure.

Earlier in the summer Will had written to Kate about the role he had conceived for Harold:

Agate, Neb.
June 8, 1922

I got your second letter from Gondola Point, when I came back last night from a trip to Scotts Bluff, about fifty miles south of here, where we parted with the Clements and Dr. Chaney who went on to Denver, while Harold and Mrs. Cook and I came back to Agate. We spent the night at Scotts Bluff and got back here late in the afternoon. There is a magnificent badland series at Scotts Bluff, not nearly so much explored as the South Dakota badlands, and more important in some ways. I am working up a scheme for an extended campaign over the country from south of there to Agate and northward with Harold to take charge of it if we can disentangle him from the ranch. I want to see him on the permanent staff of the Museum and taking charge of the collecting work here. If the Trustees and President Osborn see their way to doing that, it is going to be difficult for Harold to dispose of the ranch, but I think this is the time to try to arrange it. Most anyone can run a ranch— or if they can't it don't concern me. But Harold has a unique combination of geological and paleontological knowledge of this country that it would take me ten years to acquire with four to six months in the field. The work ought to be done and we ought to do it, but if it is possible to turn Harold in on it it will be a saving of all this preliminary knowledge that anyone else would have to acquire, but which he has already. Then also he is a brilliantly clever chap, his wife is equally clever, and is a daughter of Professor Barbour of the University of Nebraska, who is a strong force in state affairs and has a wide knowledge of the collecting localities in the state, and great collections. I want to tie in all this to the Museum; and I am sure that this is the time right now, to put for it. Times are pretty hard for ranchmen and farmers and the modest salary that we could offer would look a lot more comforting to them now than a few years ago, or perhaps in a few years to come. They haven't any servants now, have to do all their own work, and though they don't talk about such things, I think they are having pretty hard sledding. So there is my present daydream, and I hope it comes out. If it does, and the Cooks spend the winter in New York, you'll find them most congenial friends. But I expect it to work out to their living mostly out here, and getting a lot of the preparation work done here, another old dream of mine, for which I think the time is right to make another push.

One may wonder if Harold would ever have fitted into the dream—should it have materialized. He was a complex person with competing interests that almost certainly would have impinged on any pursuits he might have attempted to follow as a museum paleontologist, running a semi-independent operation in northwestern Nebraska. Indeed, the attempt to balance strong professional interests against domestic matters that constantly pulled at him was the story of his adult life.

Harold was devoted to the ranch—and no wonder. It was his beloved boyhood home, and beyond that, it was a lovely place to live. So he spent much time and energy as a young and middle-aged man in trying to maintain and improve the ranch. "Most anyone can run a ranch—or if they can't it don't concern me," wrote Will to Kate, but Will and Harold would inevitably have collided head-on over that statement. If Harold had been a museum staff member, and if there had been a ranch crisis, museum affairs would almost certainly have been put aside. Furthermore, Harold was a very independent person, as would be expected in one who had grown up as he had; he would not have accepted with equanimity institutional decisions that he did not like.

Harold was on the staff of the Denver Museum for some years, and it was not an easy time. Eventually, he became very involved in petroleum geology, and this diverted his attention from fossils. On more than one occasion he told me that just as soon as he had finished his work on certain oil wells he was going to get back to studying fossils, to complete various paleontological projects that he had in mind. He never did.

In the end, the ranch became a national monument, in which capacity it serves not only to enlighten visitors about the Tertiary mammals of the high plains but also to memorialize the Cook family, particularly Captain Cook, thus preserving a vital segment of the history of the West. Even if Matthew's dream did not come true, his vision has been fulfilled in a different way. Agate is a paleontological center under National Park Service supervision, and, as such, will be properly maintained for the foreseeable future.

The work of Matthew, Thomson, and Cook at Agate and at Snake Creek has been immortalized by the magnificent slab that Bill exhumed from the quarry, by skeletons from Agate also to be seen in New York, and by the three Matthew publications describing the Snake Creek fossils, all published in the *Bulletin of the American Museum of Natural History*. They are: "A Pliocene Fauna from Western Nebraska" (with H. J. Cook, 1909); "Contributions to the Snake Creek Fauna" (1918): and "Third Contribution to the Snake Creek Fauna" (1924).

In his third contribution Matthew suggests that the "Snake Creek beds are channel and floodplain deposits of the North Platte River at a time when the valley was at a considerably lower level than now."

Matthew visualized earlier broad sheets and lenses of floodplain deposits that were variously overlapping as the result of ancient streams
shifting back and forth. These floodplain beds, he went on to say, were
in turn cut by river channels.

> At certain points, probably due directly or indirectly to the exis-
> tence of persistent pools or springs that served as waterholes for the
> animals, fossil bones and teeth are abundant but nearly always dis-
> sociated, much broken and often waterworn. They are much scar-
> cer in the backwater of floodplain deposits but frequently asso-
> ciated, often articulated and more or less complete skeletons,
> although in many cases considerably damaged by subaerial weath-
> ering before burial.[2]

Here one is afforded a glimpse into the past of six million years
ago, when the high plains were not so high, and the landscape was
inhabited by three-toed horses, rhinoceroses, peccaries, camels, primitive deer and antelope, ground sloths, alligators, tortoises, and a host
of predatory carnivores.

[2] W. D. Matthew, "Third Contribution to the Snake Creek Fauna," *Bulletin of the American Museum of Natural History,* 50 (1924):62.

CHAPTER 17

Climate and Evolution

*W*illiam Diller Matthew was a paleontologist's paleontologist. He went into the field and diligently searched for fossils and at the same time made detailed stratigraphic studies that gave significant time-related information on the fossils with which he was concerned. He carried on elegant research in the laboratory, and he published the results of his fieldwork and research studies as scientific monographs and shorter papers. He had a no-nonsense approach to his work.

Yet he was not an isolated scientist in an ivory tower. He reached out to the public, particularly by means of some seventy-five short articles published in the *American Museum Journal* (which, after 1918, became *Natural History* magazine) that described new fossil finds, new exhibits, and other matters related to his research. He also published some American Museum Guide leaflets—each of fifty or sixty pages—on specific subjects, such as the evolution of the horse. And he published a few articles in some general magazines of the time, such as the *Mentor*—a journal long since extinct.

He did not write books for students or for the lay public, as did some of his contemporaries. Osborn was the author of a number of rather widely read books, such as his *Age of Mammals* and his *Men of the Old Stone Age*. Scott wrote *A History of Mammals in the Western*

Hemisphere—a subject that Matthew could have elucidated brilliantly in a book. And Matthew's old friend, Richard Swann Lull of Yale, published *Organic Evolution,* a college text that was a classic for many years. Yet Matthew did not choose to write such books, with one exception. One book, seldom noticed by modern paleontologists, entitled *The Science-History of the Universe, Volume VI: Zoology and Botany,* by Dr. Wm. D. Matthew and Marion E. Latham, was published in 1909 and reprinted in 1917.[1] It was part of a series of ten volumes published by the Current Literature Publishing Company of New York. This volume of 320 pages, as evident from its title, is in two parts; Matthew's part on zoology occupies the first 194 pages of the book. It is not surprising that the section on mammals fills fifty-nine pages—almost one-third of his discussion of zoology, yet he included all animals, from lower invertebrates to man. Moreover, he included a chapter on the development (i.e., evolution) and distribution of animal life. This partial book was perhaps a minor item in Matthew's published contributions, yet within its pages are passing remarks that embody some of his scientific philosophy and now and then reveal his sly humor. He begins his discussion of the evolution of the horse thus: " 'The Horse,' wrote the small boy in his essay, 'is a square animal with a leg at each corner' "—the definition obviously delighted him.

In 1915 Matthew wrote an extensive paper of about 150 printed pages, entitled "Climate and Evolution" published in the *Annals of the New York Academy of Sciences.* This extended essay, illustrated by maps and charts designed and executed by Matthew, attracted almost immediate attention in the paleontological-zoological community, and it became probably his most widely known publication. Years later, in 1939, it was reissued as a hardcover book by the New York Academy, with a preface by W. K. Gregory, a contribution by Thomas Barbour (long-time director of the Museum of Comparative Zoology at Harvard), four supplementary papers by Matthew dealing with mammalian distributions through time, and an annotated bibliography of Matthew's work by C. L. Camp and V. L. VanderHoof, which is reproduced in the present volume. The publication of this book, almost a decade after Matthew's death, was a fitting and appropriate

[1] The reprinted volume (New York: The Current Literature Publishing Co.) bears the date 1917 on the title page. A few pages beyond, opposite the table of contents, is the statement "Copyright, 1916."

tribute to his lifelong studies of vertebrate (particularly mammalian) evolution and distribution.

"Climate and Evolution" is an exposition of Matthew's evolutionary views and his theories of vertebrate evolution. To understand the significance of Matthew's researches it is helpful to examine "Climate and Evolution," although some of the matters to be discussed have been briefly touched upon earlier.

"Climate and Evolution" reveals two basic premises underlying Matthew's thinking, namely, that the continents and the ocean basins have been permanent features of the earth through geologic time, and that the migratory movements of land animals between the continents have been along inter-continental connections and near connections still present in our modern world—a Bering route between the Old and New worlds, a Panamanian route between North and South America, the two routes at the east and west extremities of the Mediterranean Sea between Africa and Eurasia, and a route of some sort through the East Indies from southeast Asia to Australia. Today, according to sophisticated modern knowledge of plate tectonics ("continental drift"), Matthew's basic premises are seen to be flawed. But we must judge his work in context. It was to be almost a half-century after "Climate and Evolution" was written before the concept of mobile earth masses became firmly established—by means of recent geophysical techniques and paleontological discoveries that Matthew was never to know. Furthermore, the greater part of Matthew's thinking was based upon his knowledge of Cenozoic mammals; in that context his placements of the Cenozoic continents and of the movements of mammals between them are mostly valid.

It should also be emphasized that Matthew was a product of his time. In 1915 the theory of continental drift, which had been proposed by Wegener only a few years previously, was viewed with overwhelming skepticism by geologists around the world—with good reason, because Wegener at the time did not have much hard evidence to back up his claims. Such evidence would come many years later. The concept of a great southern continent, as proposed late in the nineteenth century by the Austrian geologist Eduard Suess and named by him Gondwanaland, was quite familiar to Matthew. This ancient continent, which was supposed to have embraced peninsular India, Africa, Madagascar, and, by extension, South America and Australia, was founded in part on the presence of Paleozoic plants known as *Glossopteris* in the several southern continents. In its expanded form, Gondwanaland was supposed to have been a great transverse southern supercontinent, parts of which foundered into the ocean basins to leave the remnants that are today's present southern continents. Matthew, like most of the geologists and paleontologists of his day, simply did not accept Gondwanaland. For one thing, there was no evidence of the foundering of continental masses into oceanic basins, and, for another, the distributions of fossil vertebrates could all be explained by movements of animals along the existing connections and near connections.

The opposition to a Gondwanaland was especially strong among students of fossil mammals—in part because their evidence, based upon the distributions of Cenozoic mammals, accorded nicely with the present arrangement of the continents, and in part because much of the fossil evidence for a supposed Gondwanaland was based on Paleozoic and Mesozoic plants and animals, of which many mammalian paleontologists had limited knowledge. One may wonder what Matthew's position might have been had he lived to see the great geological revolution of plate tectonics that established beyond question the former close ligations between what are now our modern continents. There was not only a southern supercontinent of Gondwanaland (or Gondwana as it is more properly designated) but also a northern supercontinent of Laurasia, the two land masses constituting the all-inclusive supercontinent of Pangaea.

If he had lived to be a very old man he would have observed the studies in various geophysical, geological, and paleontological disciplines that have proved the validity of Pangaea = Laurasia + Gondwana. Would he have accepted this new concept of the earth, even though it went against a part of the thesis upon which "Climate and Evolution" was based? Perhaps.

Interestingly, George Gaylord Simpson—who succeeded Matthew and who was strongly committed to the view that mammalian distribution through time had been determined by the present position of continents and continental connections—had a hard time coming around to the modern plate tectonic concept. Eventually he did, but one senses that he did so grudgingly.

Matthew's thesis may be paraphrased as follows:

1. Climate change has been important in the evolution of land vertebrates, and is the principal cause of their present distribution.
2. The principal lines of migration in later geological epochs have been radial from Holarctic centers of dispersal. (Holarctic refers to Holarctica, the modern zoological province that includes North America and Eurasia north of the Himalaya Mountains.)
3. Geographic changes required to explain the present distribution of land vertebrates were not extensive, and for the most part did not affect the permanence of ocean basins.
4. The theories of the alternation of moist and uniform climates, with arid and zonal climates, are in accord with the evolution of land vertebrates.
5. The numerous land bridges in temperate, tropical and southern regions, created hypothetically to connect continents now separated by deep oceans, are improbable and unnecessary to explain geographic distributions. Known facts point to the general permanency of continental outlines as we know them during the later epochs of geologic time.

A unifying theoretical thread that runs through "Climate and Evolution" is the second item enumerated above. He looked at the earth as seen in a north polar projection of the modern continents, with the North Pole as the central point, from which North and South America radiated in one direction, Eurasia, the East Indies, and Australia in the opposite direction, with Africa extending down from western Eurasia. He then presented a series of identical north polar projection maps, and on them he indicated the Holarctic centers of dispersal for various groups of mammals, for the crocodilians and for the frogs, showing in each case their outward radiations from such centers. Antarctica does not appear on any of these maps.

Such a presentation, while out of step with our modern knowledge of plate tectonics, was nonetheless a reasonably accurate depiction of the distributions of many Cenozoic mammals. By the advent of Cenozoic time ancient Pangaea had fragmented to such a point that the continents as we know them were beginning to take shape, and by the middle of the Cenozoic era they were approaching the positions that they occupy today. In this Cenozoic world many of the movements of mammals between the continental masses were necessarily along the connections that persist today.

Of course, in the light of modern knowledge, some of Matthew's concepts for the radiation of mammals into their modern habitats were erroneous. Perhaps one of the most striking examples of how modern discoveries have changed earlier ideas concerning the intercontinental movements of mammals has to do with Australia's marsupials—the pouched mammals such as the opossums, kangaroos, and wombats. Matthew argued in no uncertain terms that the resemblance of certain fossil marsupials in South America to recent marsupials in Australia was not a matter of close relationship but rather of parallel descent from common ancestry. Furthermore, he definitely opposed the notion of an Antarctic connection as a migration route between South America and Australia. "The near resemblance between the modern Australian *Thylacinus* [the so-called Tasmanian wolf] and the Borhyaenidae [carnivorous marsupials of Tertiary South America] has been used as an argument for an Antarctic connection between the two. Such a hypothesis will not bear close examination."[2] But in recent years Eocene marsupial remains *have* been found in Antarctica, and they represent an animal very closely related to a fossil marsupial in Argentina. So Antarctica *did* form a bridge between South America and Australia at the beginning of Cenozoic time. (Although Matthew denied an Antarctic crossing to Australia, he did not commit himself as to how the marsupials got there.)

Such discrepancies should not blind us to the value of "Climate and Evolution" as a landmark contribution in the study of mammalian evolution and distribution. New discoveries frequently alter established views that previously seemed irrefutable.

The importance of the book is that it set forth in closely argued discussions many of Matthew's conclusions concerning evolution and the distributions of land-based vertebrates, especially mammals. He recognized and defined the importance of environmental factors in the evolution of the mammals, arguing that external environmental changes were principal agents of evolutionary change.

In advancing this argument he followed strict Darwinian principles; he was a firm believer in the variability of organisms and the forces of natural selection acting on random variations. For example, he saw a close correlation between the evolution of the grasses of the

---

[2] W. D. Matthew, "Climate and Evolution," *Annals of the New York Academy of Sciences,* 24 (1915):96.

high plains and the evolution of the horse. His interpretation of the radiation of mammals from Holarctic centers of origin into outlying parts of the earth, although too strict in some instances, did stress the movement of mammals toward climatically agreeable environments, thereby leading to wide "adaptive radiation," i.e. evolutionary development along divergent paths of similar animals living in differing environments.

He saw evolution as a bushlike branching from an ancestral stem, not as a straight-line, almost predetermined march toward a distant goal, the latter roughly expressing the view held by Osborn. And in that bushlike evolutionary growth of any particular group of mammals, Matthew thought it was the fortuitous interactions of climate, topography, and natural selection of random variations that determined which branches of the bush grew and which withered.

He rejected completely the "typological concept"—the comparison of various fossils to more or less idealized types in the determination of relationships—a concept that was widely embraced by many paleontologists and still is by some. As for the distributions of land-living animals he destroyed putative land bridges by the score—those hypothetical connections, sometimes thousands of miles in length, that were posited by some students to explain the presences of various animals hither and yon. (Many of the extreme land bridges were "constructed" by zoologists who lacked an appreciation of the facts of geology.) Matthew saw Cenozoic mammals within the perspective of geology, and it was this view that influenced his views of evolution.

Important as it was at the time, "Climate and Evolution" has lost some of its significance through the years, largely as a result of discoveries in geology and paleontology that refuted some of its conclusions. Such is a common fate of philosophical scientific works. They serve their purpose, but as the years pass they almost inevitably become victims of new knowledge. Consequently a scientific author is generally known to posterity through his or her solid scientific research, the published results of which may not make popular reading but remain viable and important for decades and more.

It was thus with Matthew. Today he is honored, even revered, by paleontologists throughout the world for his outstanding studies of Cenozoic mammals, especially the mammals of North America. *Fossil Mammals of the Tertiary of North-eastern Colorado; The Carnivora and*

Insectivora of the Bridger Basin: Middle Eocene; Third Contribution to the Snake Creek Fauna; and *Paleocene Faunas of the San Juan Basin, New Mexico,* are still paleontological landmarks, frequently consulted and studied by paleontologists in the throes of research studies on which these publications have a bearing. Indeed, paleontologists today who come into possession of one of these volumes, or others by Matthew, consider themselves the owners of a treasure.

The work of Edward Drinker Cope, Matthew's predecessor, is a case in point. Cope, as we have seen, was a man of extraordinary brilliance who became an outstanding authority in three disciplines; vertebrate paleontology, herpetology, and ichthyology. His eminence in these fields resulted from his prodigious output of some fourteen hundred publications. Included in this outpouring of scientific papers were certain works in which Cope set forth his philosophical and theoretical ideas, such as "The Origin of the Fittest," "Primary Facts of Organic Evolution," "Catagenesis, or the Consciousness of Primitive Energy," "Archaesthetic Doctrine, Or the Function of Consciousness," and so on. In some publications of this ilk Cope championed theories that may be designated as neo-Lamarckian—in other words, as espousing the efficacy of the inheritance of acquired characters—which put him at odds with his colleagues who had long since recognized the verity of Darwinian evolution. Today such works by Cope rest quietly in the wastebasket; the value of his lifework is to be found in his numerous research papers, still valid and important almost a century after his death.

This is as it should be. Once in a great while there appears a stupendous thinker, whose theoretical writings are of such momentous import that they may be timeless—Euclid, Newton, or Darwin, for example. Today Darwin's basic concept of evolution through natural selection is so firmly established that, excepting the fundamentalists, it is no longer regarded as theory but as scientific fact on which modern biology and paleobiology are founded. But it would be interesting to know how many contemporary biologists and paleontologists have read his *Origin of Species.*

Certainly Matthew did, and Darwinian evolution was Matthew's guiding star throughout his professional life. This he made clear on more than one occasion.

Preeminent among the laws which govern the architecture of
our world of life is evolution. . . . To the paleontologist. . . . evolution appears not as a theory but as a fact of record. He does not
and cannot doubt the gradual development of diversely specialized
races from a common ancestral stock through a long series of intermediate gradations, for he has before him all these stages in the
evolution of the race preserved as fossils, each in its appropriate
place in the successive strata of a geologic period. . . . Concerning
the causes and methods of this evolutionary process he finds wide
room for discussion; but of the fact, of the actuality of it, he can
have no doubt. [3]

The "causes and methods" of evolution were argued by Matthew
in "Climate and Evolution," as they are being argued today by paleontologists all over the world. In his day Matthew added his considerable
talents to the quest for a solution to the great mystery (for there is
much that we still do not understand) of how life on the earth arose
and became what it is today. The task of refining our understanding
of evolution goes on and will go on for as long as man is a curious,
reflective animal.

[3] W. D. Matthew, "The Value of Paleontology," *Natural History,* 25 (1925):267–68.

CHAPTER 18

Return to Europe

*T*he years of the First World War were stressful for the Matthews. As Canadians they were caught up right from the beginning in the emotions of the great conflict, and even though Will and several of his siblings—Bess, George, Harrison, Charles, and Jack—were living in the United States, they could not remain mentally neutral, as could many of their American friends and relatives. They were Loyalists still, and they felt the threat to the Empire—a sinister cloud that hung over their daily lives.

The tragedy of war struck the Matthew family in a very personal way, for Robert—or Robin as he was known to the family—the one Matthew son who had remained at home, enlisted in the Victoria Rifles of Canada, attained the rank of lieutenant, and, on August 12, 1916, was killed in action at Ypres. A brass plaque on the wall of the Anglican church in Saint John, New Brunswick, honors Robert Matthew, who died "for king and country," and this memorial is today a place of pilgrimage for members of the family who visit that city.

Of course the Matthews living south of Canada could hardly think of joining the fighting; they were tied to their homes by family responsibilities. Yet when the United States joined the conflict Will Matthew joined the Home Guard, an adjunct to the armed forces organized to cope with possible disturbances on the home front. Their services were for the most part not required, but they were ready, willing, and able. A photograph shows Will Matthew, the lifelong Canadian, in a regulation United States uniform of those days, stiffly standing at attention, his rifle held at the ready position; beside him are Kate in a nurse's white uniform and Elizabeth in the garb of a Girl Scout. They were prepared to, in the vernacular of the day, "do their bit" for country, if not exactly for king.

As discussed in chapter 11, the late years of World War One were difficult for Will and Kate Matthew, particularly because the entire family suffered during the great influenza epidemic of the winter of 1917–18, so much so that their new baby, Christina, died the following summer.

With the advent of the 1920s Will and Kate made plans for a trip to Europe, similar to the honeymoon trip they had not been able to enjoy fifteen years earlier. The children were now old enough to stay with relatives, in the Matthew tradition of extended family, so Will and Kate were free to go abroad for a prolonged vacation cum paleontological study trip—a nice compromise since Will was not going to Europe without making the most of nearby fossil collections, and Kate, who years before had decided that keeping up-to-date on Will's research was not on her lifetime agenda, had decided that on this trip, even though it might be a delayed honeymoon, she was not going to twiddle her thumbs while Will was contemplating fossils.

Kate wanted a sight-seeing partner, and that is why the trip to Europe was a trio instead of a duo, the third person being Mrs. Malvina Gould, a friend and neighbor. To the Matthews she was Bene (pronounced "Beany"), and to Kate she was a companion who could be counted on to share every adventure, good and bad, with the best of spirits. Will good-naturedly agreed to this arrangement; for him it presented no problems.

The departure was scheduled to take place at two o'clock in the afternoon of August 29, 1920, aboard the Swedish-American Line's *Drottingholm,* which was docked at a pier at the foot of fifty-fifth Street, in Manhattan. It was an exciting day for the considerable contingent of relatives and friends assembled to bid good-bye to the travelers. Will Matthew went to the dock early to search out the Matthew trunk and fix a proper label on it. Will Pomeroy, a friend of long standing, went to the Lee home in Brooklyn—where there had been a farewell visit—to escort Kate and the youngest Matthew, Roy (William Pomeroy, then six, later known as Bill) to the ship. It was a journey by taxi, then subway, and taxi again.

At the pier Will was waiting with his daughter Margaret, his brothers Harrison and George, Harrison's wife and son, Cama and Pete, Eloise (the wife of Will's brother Charlie), and Bene Gould. (Margaret was to stay first with the Harrison Matthew family in White Plains, and subsequently with Alfred and Marian Diller in Montclair, New Jersey; Roy with Frank and Jean Diller in Brooklyn.) There were hurried adieus, and then, at the last moment, it was discovered that Roy's bag, packed with clothes he was to wear while living with his Diller relatives in Brooklyn, had gone aboard ship along with his parents' luggage. The "all aboard" signal had been given, no one was allowed to go from ship to dock (or the other way around), and everyone had visions of little Roy's garments enroute to Europe—unused—and of Roy on the Diller doorstep in Brooklyn, a soon-to-be naked waif. However, Will was finally successful in passing Roy's bag across the gap between ship and shore and into Will Pomeroy's hands.

Like Will's first trip to Europe in 1900, this Atlantic crossing was a protracted trip, lasting ten days. Perhaps the *Drottingholm* was a faster ship than the *Maasdam,* the vessel in which Will had made the earlier crossing, but she went the long way around. It was a northern route, around the northern tip of Scotland, and from there across the North Sea to Göteborg, Sweden.

The voyage was, on the whole, uneventful, but for Bene the first night out may have been one to remember. She was assigned to a cabin that she shared with a large Swedish woman who spoke no English. On that first night Bene discovered she had (in Kate's words) "another inmate in her cabin. She thought it was a rabbit but found by using her little search light that it was a large rat. Said rat had a regular beat, she says, which he kept up all night; around the moulding, down and across her feet, and up the other moulding, around, down across her feet and so on. She seems perfectly calm about it. Says she slept fine after she found out what it was. The steward has promised a rattrap" (Kate Matthew's diary). Bene was obviously an unflappable traveler.

After landing in Göteborg they took a train to Stockholm, where they began their travels. Their manner of travel was the antithesis of that of the standard tourist; their progress through Europe was a series of mutual and separate journeys, of partings and of reunions, of shared and independent adventures, of planned and spontaneous visits. It is impossible, three-quarters of a century later, to reconstruct the complicated courses of their journeyings. Will had a plan for visiting various paleontological collections that took him from Stockholm to Berlin, Frankfort, Heidelberg, Stuttgart, Tübingen, Munich, Vienna, Florence, Rome, Zürich, Basle, Geneva, Paris, Brussels, and then to London. While he was in these various cities Kate and Bene were flitting back and forth across the European countryside, exploring fjords, rivers, castles, villages, and other places that appealed to them, but always managing to rejoin Will, wherever he was. They traveled with him between cities, but, once he was established at some paleontological museum, off they would go.

Little advance planning of these side trips occurred, and consequently Kate and Bene had their little adventures, such as getting visas and other official documents on the spur of the moment. Travel in those days was seemingly more difficult than today. For example, all three cooled their heels for hours in Sweden, trying to get visas to enter Estonia; in the end they had to give up. Later, in Italy, they "spent the whole rainy day trying to get our passports vised by the Swiss and French bureaus, and permission from the police to *leave* Italy. Failed in every thing except obtaining the Swiss visa and getting ourselves drenched" (Kate's diary).

The spontaneity of the side trips made for problems when the two ladies tried to obtain bed and board. For example, at the beginning of this European peregrination Kate and Bene decided to explore the Norwegian fjords while Will remained in Stockholm. Kate made an entry in her diary.

> We went from the station to the Grand Hotel and found they had no rooms. Left our luggage there and started out to hunt. In despair went to Cooks agency and after telephoning for some time he gave us a hotel that would let us have perhaps one room with two beds, but when we got there they said "oh no, they had not promised anything but they might have something at four but could not promise." So Bene and I staggered around till we made

up our minds we would *have* to have something to eat, and
stumbled into a little cafe and asked two men if it was a place for
ladies. When we found they spoke English we told them all our
troubles, and one of them started to find us a hotel. He spent one
and a half hours telephoning without pause before he finally found
a place where they could let us have a room at six p.m. So we
thanked him with deepest gratitude, gathered our bags together,
got a taxi, and started for the hotel. We kept expecting the driver
to stop every minute, but no, one by one the houses sped by us till
we were way out of town, and on he went. At every possible place
we would say I wonder if this is it? At last we drove up to a sort of
summer hotel and heaven be praised they took us in.

In the meantime Will Matthew was quietly studying fossils in
Stockholm.

This pattern of activities—busy and frenetic on the one hand,
quiet and studious on the other—was typical of the Matthew-
Matthew-Gould journey up and down the face of Europe during the
fall of 1920.

It is no surprise that doors were open to Will Matthew wherever
he went. No longer the young novice of 1900, Dr. Matthew was now
an established paleontologist of worldwide repute whose opinions
were eagerly sought in museums and universities along his travel
route. He undoubtedly had written to inform various paleontological
colleagues of his plans. Certainly this was true at Tübingen, in southern Germany, where Baron Professor Friederich von Huene and his
family were eagerly awaiting the arrival of the Matthews.

Von Huene was an old friend. He had studied at the American
Museum of Natural History before the First World War and at that
time had gone on a field trip in the Southwest with American colleagues. He came from an old aristocratic German family but spoke
English fluently. Von Huene was a dedicated paleontologist, a devout,
unworldly person who devoted his long life to the collection of and
research on fossil vertebrates, particularly reptiles. It is likely that he
had much to talk about with Will Matthew. Moreover, he was a
courtly gentleman of the old school, and went out of his way to be
helpful to Kate and Bene. Kate noted in her diary that

we were totally unprepared for the enthusiastic welcome he gave
us when we reached his Museum at the University at ten a.m. He

was lovely about taking Bene and me to an office where we had our time in Germany extended on our passports, and then after a casual survey of his collections he took us all up to his house where we met his wife and their five little daughters. They are named Erika, Olga, Alexandra, Irmgard and Ursula, and are quite proud of having all the vowels of the alphabet represented [the initial letters of their given names.] We had a large and ample dinner but very simple, no meat, and black bread, with apples for dessert.

The hospitality and the simple meal were quite typical of Baron von Huene and his family. Although more straitlaced and ascetic than Matthew (von Huene abhorred smoking and liquor, while Will Matthew smoked vile cigars and enjoyed his daily glass of beer), the two men must have relished their time together, for they did have much in common. Not only were they dedicated paleontologists, they were both terrific walkers; a long hike was the height of enjoyment for Matthew and for von Huene. They probably had some good walks together in the hills around Tübingen. They spent a day together at the Trossingen dinosaur quarry, where at the time von Huene had a large crew excavating skeletons of the Upper Triassic prosauropod dinosaur, *Plateosaurus.* Indeed, Professor Osborn had arranged financial aid for this excavation, and in the end a very fine skeleton, one of several that came out of the quarry, went to New York, for exhibition. (It can ordinarily be seen in one of the dinosaur halls at the American Museum of Natural History, although these halls are presently closed for renovation.) Matthew thus had an immediate interest in making a close inspection of the quarry, and what was coming to light as a result of the digging there.

The journey continued through southern Germany and into Austria, with the same pattern of activities—Will studying in museums while Kate and Bene explored museums and galleries in the cities, and castles and villages in the countryside. One short paragraph in Kate's diary, written in Munich, explains it all. "We got back to the hotel half dead at 5 p.m. and found Billy here. He had had a most successful time and all the scientists had been *so* nice to him." And noting the occasional exception to the "rule" that called for them to spend their days apart, Kate wrote in Vienna that "it was *so* nice to have Billy with us for once. And he did enjoy it so."

The trio went on to Venice, where together they behaved more in the fashion of tourists than had been the case during their travels through Germany and Austria. In Rome and Naples they saw the sights—the Forum, Vesuvius, and Pompeii, as proper tourists should. But for the remainder of their trip on the Continent they reverted to the established routine whereby Matthew spent his days in paleontological museums while Kate and Bene were strenuously visiting art galleries and castles.

Will Matthew made and renewed many valuable paleontological contacts: Ferdinand Broili and Stromer von Reichenbach in Munich, Othenio Abel in Vienna, Hans Stehlin in Basle, Marcellin Boule in Paris, and Louis Dollo in Brussels. He made notes on fossil collections and discussed paleontological and evolutionary problems with the paleontologists of Europe, all of whom were happy to be his host. These scientists, for whom national boundaries were—and still are—artificial obstructions to the free flow of knowledge, had been separated from many of their colleagues during four violent years. Matthew, one of the first North American paleontologists to visit them, brought with him fresh, new ideas from that part of the world.

While Will was making his final study in Brussels, Kate and Bene were walking across the old battlefield of Waterloo, and the recent battlefield of Ypres, where destruction was to be seen on all sides. Then, on November 24, Kate and Bene "met Billy on the train and all came to Ostend together," where they took the channel steamer for Dover. It was close timing.

The final two weeks of their European trip were spent in England, where (among others) they visited Arthur Smith Woodward, the doyen of British vertebrate paleontologists, who four years previously had written the letter to Will Matthew offering to sponsor him for fellowship in the Royal Society. They also were guests of the Elliot Smiths (Elliot Smith was one of the leading students of man's evolution as revealed by fossils), of the D. M. S. Watsons (Watson was then in the early stages of a most eminent paleontological career—he was a student of fossil reptiles, and in his later years a Trustee of the British Museum), and of the Forster-Coopers (Clive Forster-Cooper had studied at Columbia under Osborn before the First World War and subsequently was a professor at Cambridge and then director of the Natural History Division of the British Museum.) These visits were short intervals in their usual schedules, which saw Will Matthew

continuing his careful studies of fossils, principally at the British Museum, while Kate and Bene were making the rounds of the National
Gallery, the Tower of London, Windsor Castle, Canterbury, Cambridge, Bath, Stonehenge, Amesbury, Salisbury Cathedral, Winchester
Cathedral, and many lesser points of interest. It was a very full schedule, but they managed it and got to Southampton on the night of
December tenth where, on the next day, as Kate wrote, "we met Billy
at the dock alright."

At midnight they sailed on the ship *Imperator* for New York.

CHAPTER 19

Travels in Asia

*D*uring the years following the First World War, including that eventful summer of 1920, when Will, Kate, and Bene Gould traveled up and down Europe, plans were being made at the American Museum of Natural History for a series of large and elaborate summer campaigns into central Asia in search of fossils. The impetus for this program came initially from two members of the museum family, Professor Osborn and Roy Chapman Andrews. For Osborn it was to be a quest to find "early man," because he visualized this unexplored, and at the time largely unknown, hinterland deep within the vast Asiatic continent as the ultimate birthplace of humanity. (Years later it was to be revealed that the origin of human beings took place in southern Africa, not in Asia.) For Andrews it was to be the culmination of a dream of exploration on a grand scale.

In those days such an ambitious undertaking posed formidable problems for Osborn and for Andrews, as well as for other people who were to become deeply involved in making this projected incursion into the northern deserts of Asia an actuality. The early twenties predated by decades governmental support for large scientific projects, particularly such support as embodied today in the National Science Foundation. All money had to be secured from private funds, and the formidable task of raising such funds involved month after month of pleading, cajoling, and begging, especially on the part of Roy Andrews. In the end the efforts of dedicated parties at and in support of the museum were successful, and what was eventually to be known as the Central Asiatic Expedition came into being.

There were five successful expeditions: in 1922, 1923, 1925, 1928, and 1930. And an aborted expedition occurred in 1926—the expedition that relates to the story of Matthew and his life.

The three earlier expeditions were paleontological and public triumphs of the first magnitude. No early humans were found, but in the sediments of Mongolia Andrews and his cohorts found early mammals of all kinds, dinosaurs, and dinosaur eggs. The dinosaur eggs, in particular, caught the public imagination and made the Central Asiatic Expedition world famous. Perhaps the eggs were not the most important fossils uncovered in Mongolia, but they tickled the fancies of thousands of people who read newspapers and magazines or who listened to primitive radio programs, and that was all to the good. The CAE (as the series of expeditions were commonly designated by the paleontological fraternity) had achieved a solid record of accomplishment in the scientific world and in the world at large.

The chief paleontologist of the first three expeditions was Will Matthew's lifelong friend, Walter Granger. It was the wish of Granger, as well as of Osborn and Andrews, that Dr. Matthew should be a member of the 1926 expedition—to see the field relationships of the locales where in previous years Granger and other members of the expedition had collected the fossils, especially the fossil mammals, that had in many instances been thereafter studied and described by Matthew. So it was that in the autumn and winter of 1925–26, plans were made for Dr. Matthew to participate in the forthcoming summer expedition.

Once again, as in 1920, the plan was to include Kate and a friend—this time Virginia Fulenwider. Will was to leave in March and sail from Vancouver to China, subsequently to explore Mongolia with Andrews and Granger and other members of the expedition. Kate and Mrs. Fulenwider were to join Will Matthew in Peking in the late summer, and then they were to travel more or less together to India via the East Indies. Kate and Will would then go on to London.

It was Matthew's desire, after the completion of his explorations in

Mongolia, to study fossil mammals from the Siwalik beds of India, some housed in the Indian Museum in Calcutta, some in the British Museum (Natural History) in London. It should be explained that in 1922 Barnum Brown had made a large collection of fossils from the Siwaliks for the American Museum. Matthew was the logical person to study and describe this collection. The Siwaliks, incidentally were referred to in Rudyard Kipling's great story *Kim*. Kim and his companions, Hurree Babu and the lama, "crossed the Sewaliks" on their way into the high ranges of the Himalayas—the "Sewaliks" being the foothills which to the south form the eaves of the roof of the world that separates the Indian subcontinent from the rest of Asia.

Matthew's Asian venture proved to be an extended trip, lasting from March 1926 to the following March. For Kate the trip extended from the late summer until its predetermined end. By now the Matthews felt that the children were far enough along on their separate paths toward adulthood so that such a parental absence would not be deleterious. In fact, Elizabeth was to graduate from college in the spring of 1926 and was becoming quite an independent person. Margaret and Bill were to spend the academic year 1926–27 with the Charlie Matthews in Portsmouth, Virginia—Margaret as a high school senior, Bill as a junior high school student. It would all work out very well, thanks to the Matthew practice of intrafamilial cooperation.

In mid-March of 1926 Will Matthew was on a Canadian Pacific train bound for Vancouver. From Vancouver, on March 20, he took a ship for China, and in early April was in Peking,[1] getting settled and making plans for his journey into Mongolia.

Having arrived in Peking, Will Matthew was confronted with the unpleasant and ominous fact of Chinese internecine warfare, then prevailing in northern China. Undoubtedly, he had been briefed on the situation through communications from Roy Andrews in Peking to the museum authorities in New York, but perhaps the stark realities of the situation were not so apparent until he was there on the ground, with the fighting at his back door. It was the period of the Chinese warlords, when life in the celestial empire was uncertain. In

---

[1.] Today, of course, the capital of China is called Beijing, but here the old name will be used—as it was used by Matthew and his colleagues. The same practice will be applied to other Chinese place names cited by Matthew.

a letter to Kate in Hastings-on-Hudson, written on April 25, Will described some events in China.

> Still in Peking and no immediate prospect of getting off. The Kuominchun have retreated to Nankou Pass between here and Kalgan [Chang-chia-Kou], and the "Allies" (i.e., Chang-tso-lin, military chief of the Manchurian provinces, Wu-pei-fu, chief of Central China, and the Shantung chiefs from east and south of here) are said to be pursuing and trying to surround them. One good authority says that there were twenty thousand casualties on that front last week; another equally good says that there was a lull and no serious fighting last week, but probably will be a big battle this week. And so on. We can't get through until one or the other is beaten or they stop fighting at least temporarily, so don't know when we will get away. Meantime I find plenty to do with study and figuring the specimens from last summer, prepared here during the winter.
>
> By the time you get this you will be making plans for going down [to Sweetbriar College] to see Elizabeth chucked out on the wide, wide world, and you will be starting to pack things away for the summer and summer weather beginning—and I'll be out in the desert—I hope—hunting Eocene rhinoceroses in the Shara Murun district, or possibly pushing westward in a reconnoissance into the unknown country (geologically) that lies beyond . . . I wish we could plan details of the expedition more definitely, but everything is in a mess at present.

As if to emphasize what Will had written to Kate on April 25, an item in a Peking newspaper, dated May 12, described the tribulations experienced by McKenzie Young, in charge of the expedition vehicles, when he went to Kalgan in an effort to evaluate the possibilities for getting the expedition beyond Kalgan and into Mongolia.

The heading for the article read "Expedition Motor Chief returns from Kalgan.—Gets through war zone and reaches Peking after week of exhausting travel; tells Andrews suicide to take Expedition through battle lines."

The story continued:

> Peking, May 12.—After seven days on the road, J. McKenzie Young, Chief of the Motor Transport of the Central Asiatic Expedition, reached Peking yesterday morning from Kalgan.

Mr. Young started from Kalgan last Thursday by train, going as far as he could and then began to walk. He finally got to the railhead at Taiyuanfu via Tatungfu.

Not having had his clothes off for the past five days and with little sleep, Mr. Young was in an exhausted condition upon arrival yesterday morning. He was at once put to bed so that his story of what is happening in the Kalgan region was only obtained partially.

He assured Mr. Andrews that it would be nothing short of suicide to attempt to take the Expedition from Peking to Kalgan through the battle lines. The war zone is thick with snipers and it was Mr. Young's opinion that passports or other papers would be of little avail to save the party from death.

Shortly after the appearance of this newspaper story Roy Andrews gave an interview to the press in which he expressed a very discouraging view about the possibilities for summer fieldwork by the Central Asiatic Expedition.

Andrews says plans shot to pieces by war. Expedition to Mongolia to find Man's birthplace can't start till war-lords open up route.

"Our plans for this summer are all shot to pieces" said Dr. Roy Chapman Andrews, leader of the Central Asiatic Expedition yesterday morning.

"While we are ready to start from Peking tomorrow, there is nothing to guarantee us safety through the Fengtien and Kuominchun battle lines," continued Mr. Andrews.

"Negotiations have been under way with both combatant factions but the poor communication facilities with Kalgan made almost impossible the achieving of any satisfactory results," explained the leader.

Although the camel caravan got away from Kalgan in the middle of April, most of the motor cars of the Expedition as well as the foreign personnel are stuck in Peking.[2]

It should be explained that the Central Asiatic Expedition had developed a tactic, very efficient for those days, whereby a large caravan of camels would set out, loaded with drums of gasoline. The motor cars, carrying the scientific personnel and assistants would then meet

---

[2.] The quotations are from newspaper clippings in the Matthew archives. The names of the newspapers are not available.

the caravan at a predetermined rendezvous point, where gasoline drums would be transferred to the cars, while fossils and other scientific specimens would be loaded on the camels, to be transported back to Kalgan for shipment to Peking.

In April 1926 the camels were on their fruitless way, while the cars remained trapped in Peking, and that was the situation that greeted Will Matthew when he arrived at the expedition's headquarters. The outlook was grim, but he refused to give up hope.

Nonetheless, hope was day-by-day becoming ever more tenuous. On May 16 Will wrote to Kate in Hastings that "'the scripture moveth us in sundry places' with very strong and well placed words—but they ain't nothin' to the moving words we are using—quite super-scriptural words to express our opinion of this delay. The situation isn't a bit better, worse if anything, and we can neither go through nor go 'round until they either fight it out or make peace. We may still be here in June so far as present appearances go."

Yet some hope was left, as expressed in a following letter to Kate, written on May 26.

> If we don't get away within a couple of weeks I am afraid we'll have to give up the long trip into the far interior toward Turkestan, and content ourselves with getting out to Shara Murun, some 300 miles beyond Kalgan and settle down to collect in the Eocene beds there for what is left of the season. We can do that even if we can't get through the lines until fairly late in the summer, and it will be interesting collecting, in some of the faunas that I have been especially working on for the last year and would be particularly interested on getting more material and studying the stratigraphy of the formations, succession of the faunas and conditions of deposit etc.

The frustration of waiting, wearing though it might be, did not prevent Matthew from spending some of his days at interesting and fruitful pursuits. For one thing, he could hike around within the confines of Peking to see some of the sights—and hike he did, although he feared that he had "lost face entirely with the Chinese by walking instead of going about in rickshaws as most everybody does who can afford to ride. But I need the exercise, and prefer walking and looking about me at leisure."

Furthermore, in addition to studying at the museum compound in Peking fossils that had been collected on previous expeditions into

Mongolia, Matthew was able to devote some of his time to a project that had long been in his thoughts.

> I am spending my time in working on a practical textbook of fossil vertebrates, something that can be used for field identification of the fossils by geologists as well as paleontologists. Roughly, putting down on paper what I carry in my head to identify fossils when I see them. Rather different from the ordinary text book which is intended for the cultural instruction of college students. I'm not telling what a dinosaur looked like in life or why he became extinct, but how to tell a dinosaur fossil—his name when you meet him in the field. I don't know whether I can make good on it but I'm going to try. But as usual I am desperately in need of reference books and particularly of specimens, as I want to make a great many drawings of teeth, etc. to illustrate and compare the characters of the various kinds.   (letter to Kate, May 26)

Two months later, still in Peking, Will was still planning the "practical textbook," and again he wrote to Kate about it.

> Have been working on a plan for [the] practical textbook of fossil vertebrates, which will never get very far until I can make drawings from specimens, and there are no collections here in Peking to draw from, and very few books available. I am up against the same trouble as in trying to do any scientific work at Hastings. [Matthew had a comfortable study in the Hastings house—a gracious room with a fireplace, but he seldom attempted to do any writing at home.] I can't evoke a camel from my inner consciousness, like the German philosopher was said to have done. Osborn may be able to, but not I; so I can't pick some pleasant surroundings and dictate books to a secretary.... Perhaps I could write farcical nonsense to cover my ignorance at critical points.
>
> (letter to Kate, July 25)

These last two sentences were obviously directed toward Osborn's practice of composing books in the quiet solitude of his castle, high above the Hudson, opposite West Point.

This book project was something on which Matthew had already spent some effort. The late George Gaylord Simpson recalled Matthew's work on the "practical textbook" when they were together on a paleontological collecting trip in Texas, during the summer of 1924.

Our field trip together in Texas, during most of which we shared a room, naturally led to as great an intimacy as was possible between men of such different ages and attainments.

His amazing store of information & extensive knowledge of fossil mammals, which I am sure no one else has ever equalled, was impressed on me by an incident in Amarillo. There was some confusion in finances & we had to wait several days to complete purchase of the Ford [their field car]. During that time we were staying at different hotels, due to our separate arrivals. I amused myself largely by gossiping with any local person I could draw into conversation, while he kept largely to his room. Calling on him there one afternoon I found him writing what was obviously not a letter and learned that he was improving the time by writing part of a prospective book. This was to be a sort of field book of fossil mammals, to contain sufficient diagnosis to recognize any known genus. He was writing this extremely technical & detailed work without the use of any notes or reference books of any sort.

Incidentally, I do not know what became of this proposed work, which would have been very useful to all paleontologists & geologists. That he continued with it when opportunity presented I do know, for he spoke of it again several years later as still progressing.[3]

Obviously, opportunity presented itself again in 1926, but, unfortunately, the book was never completed.

It was a wearying business waiting in Peking week after week, and trying, perhaps halfheartedly, to write some text for his book—a task for which he had at the moment little enthusiasm, since he did not have the scientific literature or the specimens to back him up. Yet there was a flicker of hope remaining, and he harbored thoughts of the expedition possibly getting out to Shara Murun, west of Kalgan, to collect Eocene fossils, as he had mentioned to Kate in an earlier letter. "That will be mighty interesting to me personally, if less important for the expedition than the long trip far west that we had planned. ... If we start very late, or if we get into rich pickings we may not get back until pretty late in September" (letter to Kate, May 31).

[3] George Gaylord Simpson, "G. G. Simpson's Recollections of W. D. Matthews," *Palaios,* Vol. 1, No. 2 (1986):200–203.

By the middle of July, these hopes for even an abbreviated, less than satisfactory expedition had been dashed. He told Kate that "things have been going from bad to worse about my Mongolian trip, and at present there seems to be poor prospect of our getting out at all. So much so, that we have had a cable from the Executive Committee at the Museum, suggesting that I had better go on to Calcutta as soon as practicable, and not wait longer here. Of course I'm not going without you" (letter of July 19).

Of course he was not. Will and Kate had planned this venture months before, at Hastings, and he was not about to change that part of the plan. The original schedule had Will coming back to Peking from Mongolia, probably late in the summer, where Kate, having arrived with her companion, Virginia Fulenwider, would be waiting for him. But when these last letters were written, Kate and Virginia were already en route—the letters of July 19 and 25 had been sent to Kate in Japan—with tickets for the remainder of the trip in hand. There was no turning back at this point. So, with plans drastically revised, Will would wait in Peking for the two peregrinating ladies, instead of the other way around, and they would then continue their trip as originally scheduled.

Or so they thought. But there were to be still more changes in their itinerary—changes required by a surprise that awaited them in India, more of which later.

Of course the abandonment of the 1926 work in Mongolia affected all members of the Central Asiatic Expedition. On August 8 Will wrote to Kate that there was "nothing new as to war conditions except that Andrews has found it advisable to sell his gasoline and food supplies up in the Shara Murun district, so that there is no further prospect of even a small trip up that way. Oh damn!"

And on August 12 he wrote to Kate, still in Japan, about the disbanding of the 1926 Central Asiatic Expedition.

> Shackelford [photographer] left on his return to the States yesterday afternoon and we all came down to the station to see him off. The Grangers and Nelsons [Nels Nelson, archaeologist] are leaving in a couple of days for Yunnan. [Walter and Anna Granger spent the winter in Szechuan, where he made a magnificent collection of Pleistocene mammals from the limestone caverns of that region] and Andrews is off for the States on the 29th. . . . So Mac

Young [chief of motor transport] and Beckwith [geologist] will be the only ones to stay here through the winter.

Will would stay on in Peking for two weeks, to be on hand when Kate and Virginia arrived. They came by train from Mukden on the morning of August 26, and Will was on the station platform waiting for them. Kate and Virginia were then taken to the truly magnificent home of Harry Payne, a friend with whom Will had made arrangements for their week's sojourn in Peking. It was to be a week of sightseeing—something that Will and Kate had been looking forward to for months. It was not the somewhat leisurely sightseeing that they had expected as an addendum to Will's Mongolian expedition, but instead a very full seven days of trips and social events to be squeezed in before they left Peking.

On the morning of September 3, the trio departed from Peking by train for Hankow where they were to take a boat down the Yangtze River to Shanghai. As is evident from Kate's journal it was not a routine journey. They were approaching a war zone, and anything might happen.

> We . . . slept in our clothes all night. At 7 A.M. we were within 20 miles of Hankow and could hear faint booming of the cannon at Wuchang. We did finally pull in at the Hankow station at 1:35, making us 39½ hours late on what should have been a 36 hour trip, viz instead of 1 night and two days we were 3 days and 3 nights and over, en route. Great scramble to get across freight cars with our baggage and out into the street where we had it sent up by rickshaw. We walked—saw wire entanglements across the street and a gun and gunner behind sand bags. . . . American Consul says they have had 3 bad nights of cannonading all around here.
>
> (Kate's diary, September 6)

The next day in the afternoon they boarded the steamer, to go down the river, and then "about 6 p.m. they told us they probably wouldn't leave till 5 the next morning and my heart sank. There we had to stay out in the river with the guns banging away on the other side, the Cantonese soldiers marching into Hankow, etc. There was nothing to do but to see it through and watch the flash from the guns at Wuchang but the shells didn't come close to us as they had the day before, so the Captain said" (Kate's diary, September 7).

After the harrowing hours on the river at Hankow the two Matthews and Virginia Fulenwider had a most pleasant trip down the river to Shanghai, where they spent four days. They then boarded a ship, the *Malwa*, bound for Hong Kong. From Hong Kong they continued their voyage on the *Malwa* through the South China Sea to Singapore, where they transshipped to the steamer *Ellora*, bound for Pinang, Rangoon, and Calcutta.

Looking over the notes written more than six decades ago, one cannot help but being impressed by the time and the trouble involved in those days of surface travel, compared with the air travel of today. For the Matthews the trip from Shanghai to Calcutta was a journey of three weeks duration, involving stopovers and transshipments; today it would be a matter of hours. For the paleontologist today a trip abroad for the purpose of collecting fossils or studying in museums is an almost routine affair. For the traveler in Matthew's day (and I can remember those times quite vividly) crossing the oceans to foreign lands was a great adventure.

Finally, on October 5, the *Ellora* docked in Calcutta, where Will was to begin the second assignment of his Asiatic trip, research at the Indian Museum on fossil mammals from the Siwalik beds. But a little holiday jaunt to Darjeeling was planned first. On October 11 Will, Kate, and Virginia took a train northward where eventually they transferred to the toylike, narrow-gauge railroad that climbed, with many twists and turns through the hills, to that mountain resort where, in the days of the Raj, so many British sahibs sought relief from the searing heat of the lowlands. It was a relaxing week, and on the eighteenth of the month they were back in Calcutta, and Will was ready to begin his Siwalik studies in earnest.

The next morning he met Dr. Guy Pilgrim at the museum. Pilgrim, an official of the Indian Geological Survey, was an "old Indian hand", having spent many years in the subcontinent collecting and studying Siwalik fossils. He was eager to introduce Matthew to the collections and to assist in the many ways that paleontologists help visiting brethren become acquainted with the fossils in the host's institution. On that day it seemed as if Matthew was set for a long and fruitful visit, studying in detail the fossils that he had for so many months looked forward to seeing.

The next day, October 20, his visions of the days to come were abruptly changed by a cablegram he received from Professor Osborn. A new *Pithecanthropus* (or *Homo erectus*, according to modern usage) skull had just been discovered in Java, and Osborn was anxious for Matthew to go there and examine it. Two days later another cablegram from New York arrived; Osborn stipulated that Matthew *must* go to Java to see the new skull.

It was hard. Will had to revise his program for extended research in Calcutta and retrace the long sea voyage that he had just completed. Many weeks would be knocked out of his schedule. As Kate wrote in her journal, "why wouldn't it have happened while we were in Singapore!"

But it didn't, so now all their plans were drastically revised. It was decided that Will would leave Calcutta on November 15 for the trip back to the East Indies, while Kate and Virginia would depart on October 27 for a roundabout trip to Bombay—during the course of which they would briefly interview Mohandas Gandhi—and would see many of the famous sights in India. In Bombay Kate and Virginia would go separate ways. Kate would then go by ship to Egypt, there to wait for Will to join her in December, during which time she would travel to Palestine and make a trip from Cairo up the Nile. Will would have to miss these opportunities for sight-seeing; owing to the constraints of time he would have to sail directly from the East Indies to Cairo, where there would just be time to catch the ship that would take them to Marseilles and from there to London.

In the meantime he would make the best use of the few weeks left to him in Calcutta—spending all his time with the Siwalik fossils. It proved to be time well utilized, for, with the additional two months that he later devoted to Siwalik fossils at the British Museum, he was able to compile research that resulted in a publication of some 125 pages, with illustrations, appearing in 1929 as a *Bulletin of the American Museum of Natural History*. This paper was entitled "Critical Observations upon Siwalik Mammals (Exclusive of Proboscidea)." Matthew wrote in his introduction to the Bulletin that "the object of this study was to check up in the light of modern paleontological evidence the classic researches and descriptions of Falconer and Cautley and of Lydekker, and the admirable later work of Pilgrim, as a basis for researches and description of the collections obtained for the American Museum by Mr. Barnum Brown in 1921–1923."[4] And the objective was well achieved. Matthew's all too early death in 1930 precluded any serious work by him on the American Museum's collection of Siwalik mammals, and it transpired that the privilege of describing this collection fell to me. "Critical Observations" was invaluable in carrying out this task.

With Kate and Virginia on what proved to be an eventful trip across India, Will Matthew was very busy with his paleontological studies during those final weeks in Calcutta. He wrote to Kate about some of his experiences, which were not always pleasant.

> Green flies are nothing less than awful. I close the shutters before lighting up but none the less there is a steady shower of them beneath each light, and the floor is a shambles in the morning with windrows of corpses over its surface. Also there are mosquitoes, even more small and active than the Chinese ones were, and poisonous little devils too. . . .
>
> I don't like this town. The more I see of it the less I like it. Socially and in other ways it is rather a contrast to Peking.
>
> The fan in this room has a squeal just like a couple of hawks or kites and has waked me up at night two or three times with the impression that some of the birds had got into the room and I would have to chase them out.   (letter to Kate, October 31)

His dreams of kites invading the room were not entirely fanciful. Since it was and still is common in India, as in many tropical countries, for unscreened windows to be left open, it is not unusual for animals other than insects to invade the sanctity of the household. One may sit in a quite elegant drawing room in the evening and watch the bats flit by in their pursuit of insects attracted by the lights. There is a story of the time Dr. Guy Pilgrim was busy studying some rare fossils in his laboratory, when a mynah bird flew in and snatched a very precious fossil tooth from under his nose. The saucy bird flew into a tree nearby, followed by violent oaths and imprecations from Pilgrim, who was perhaps denied thereby the opportunity to name a new species of extinct mammal.

---

[4] W. D. Matthew, "Critical Observations upon Siwalik Mammals (Exclusive of Proboscidea)," *Bulletin of the American Museum of Natural History*, 56 (1929):437.

Matthew's sorrows were not all the result of outside influences. In the same letter he recounts an exasperating event that must have left him helplessly indignant.

> Oh yes, one very sad misfortune. I was filling my drawing pen with India ink, and somehow it slipped in my hands and splashed ink all over the front of my pongee coat. Of *course* I had to be wearing a coat, which I don't one time out of ten when at that work in the office, and of *course* it had to be the silk and not the cheap drill that I'm wearing in place of it. I never had that kind of an accident happen before, and of course never will again. But the coat is ruined, unless I can mix some paint of the right color and paint over the spots. Oh dam! And likewise potatoes!

In a letter written on November 4 Will told Kate about the progress of his work on Siwalik fossils. And, concerning specimens and past practices, he remarked "I have found specimen after specimen which I've succeeded in piecing together from fragments supposed to be of different individuals and in several cases collected in different years! And referred to different species! Pretty rotten work professionally, but I don't think Pilgrim is so much responsible for it as old Dr. Lydekker who worked on the collections some thirty or forty years ago. A very learned man but inaccurate and careless to a degree."

So much for one of the great names of British (and Indian) paleontology.

On November 7, as he was nearing the end of his Calcutta studies, Will wrote to Kate that he "had a telegram from Java saying that the finder of the new skull would permit examination of it; but that he did not consider it to be a *Pithecanthropus* skull (true 'missing link') but a species of man contemporary with the *P.* It will be interesting to see, anyway."

What was he destined to see in Java? Something new and epochal? Or something quite ordinary? Will must have entertained mixed feelings at the receipt of this news; was his trip, to be made at such cost to his original plans, going to be worth the effort?

On the day before he was to sail to Java, November 14, he wrote to Kate that he

> Got a message from the Museum yesterday, saying that Geo. Olsen [technician] would not go to Szechuan but stay in Peking and could

be available if desired for work in Java. [Evidently Professor Osborn had daydreams of Matthew unearthing a whole new series of *Pithecanthropus* remains, thus doubling, and more, the then meager record of the earliest hominids.] They don't quite get it that Java is south of the equator, that the rainy hot season is from November to March and the working season April to October. Well, I wrote the boss [Osborn] a long letter. I hope he reads it—it'd do him good but he won't.

Tomorrow morning early I leave on the *Arankola* for Rangoon, transship there to the *Egra*, and transship again at Singapore to the first boat I can get for Batavia [present-today Djakarta]; thence by railroad to Bandoeng, and there arrange a visit to Trinil [where the original *Pithecanthropus* was discovered] and other places. Trinil is in the Madoeng residency, in the centre of the island or a bit west of the centre.

Weeks later it seemed to Will Matthew that the *Pithecanthropus* skull he had come so far to see was more like a will-o'-the-wisp than a substantial fossil. When he landed in Batavia the fossil was not there—but he was made to understand that it would be available somewhere along the route from west to east that traversed much of the length of the island—the route that he had laid out for himself, as mentioned in his letter to Kate. The first stop on that route was Bandung where he was ensconced in a very comfortable hotel. Indeed, Matthew found Java to be a well-organized and well-kept country. "The whole city is clean and decent and well managed. I must say these Dutchmen know how to run a colony" (letter to Kate, undated). His letter continues:

I spent a good part of today with the Geological Survey officials, who have been most friendly. They showed me the specimens that have been collected recently and tomorrow I am going to make some detailed notes and sketches of them, and Thursday probably we start for Boemi Ajoe [now Bumiaju], a new and important locality about the centre of the island. Thence we go on further to Madioen [Madiun] residency, where there are several localities including the one where the famous missing link skull was found, and then on to Sourabaya [Surabaja], where Dr. Heberlein has the supposed new skull of the missing link (which I gather the scientific staff here are pretty skeptical about).

Dr. Heberlein was a German physician who had found the supposed skull, which was the cause of all the excitement and the entire reason for Matthew's reversal of course—back to Java from India.

From Bandung on Matthew traveled by car, the driver being Dr. van Es of the Geological Survey. Will Matthew found travel by automobile in Java most interesting: excellent roads, good hotels, tropical scenery, friendly people—all making his journey delightful and memorable. He imparts some of the exotic charm of such travel in a letter written to Kate from Madiun on December 7, 1926:

> How you would have enjoyed the trip I have been taking with Dr. van Es, traveling by auto over these splendid roads all lined with natives going to and fro or sitting around in the shade. Endless numbers of them, men, women and children, mostly on foot, but some in little two wheeled carriages or driving heavy two wheeled carts, the latter with carabao, the former with ponies. Children herding little groups of carabao along the road, or small flocks of sheep or goats, or gathering grass with a little sickle-shaped macheté. Women loaded with baskets full of fruit or corn or miscellaneous other things, or with a baby strapped on the hip or on the back. Men mostly parading along empty-handed, but often toting a couple of heavy baskets on the ends of a bamboo pole across the shoulder. . . . Nearly every native goes barefoot, only a few wear sandals.
>
> The roads are lined almost everywhere with native huts, and a village every now and then with a group of open booths, an open market place with a row of covered platforms on which the goods for sale are laid out.
>
> Dr. van Es is a skilled but very fast driver. . . . Can be done on these fine smoothly graded asphalt surfaced roads, but means a lot of traffic dodging. We normally have to walk several miles from the nearest point on the road to the localities we have to inspect, so I'm ready for bed when we get through the day. Get up at 5:30, breakfast at 6:30–7, and off before 8 o'clock. In the afternoon it is hot, and apt to be heavy tropical showers, so the morning is the best walking time.

On the way to Madiun Matthew visited the Trinil locality, where Dr. Eugene Dubois had discovered the famous *Pithecanthropus* skull and femur in 1891, thus revealing to the world the first remains to be found of truly early hominids. For Will Matthew it was something in the nature of a pilgrimage to a paleontological shrine, but one that had to be viewed from a distance for "there was nothing to see, as the river was so high in flood that the excavations were all covered up."

As Will Matthew proceeded on this visit to a land where ancient, prototypical men had lived, he was incessantly plagued by thoughts about the *Pithecanthropus* he was supposed to see and evaluate. These thoughts surface in the letters he wrote to Kate—particularly in one written on December 2 and 4 from Bumiaju:

> We go to . . . Sourabaya, where the new skull is, if the finder has not already sent it back to Batavia. I may slip on seeing it at all, as the finder may have sent it before I arrive and it may not have been unpacked at Batavia when I get there. As I have reason to believe that it is not very much of a find—not a perfect skull but a cast or interior mould of the cranium only (if it be a fossil at all), and that the finder himself is not at all sure of its being the *Pithecanthropus* or different from an ordinary human skull—and as the finder is a German doctor—and for some other reasons—I rather suspect that this is deliberately arranged with the intention of sidestepping my seeing the specimen. Better keep this to yourself. I am quite sure that the Geological Survey people are not parties to any such scheme, and are quite in the dark as to the real nature of the find. They have done everything possible to arrange for me to see it, but it has been or is being turned over to the Department of Education! by the finder, probably with the conditions expressed in their letter, which stated that I could see it in Batavia. It is all a complicated matter, difficult to explain on any other hypothesis than that Dr. Heberlein, the finder, realizes that it is not a *Pithecanthropus* skull but does not want to admit it himself or to have any experts see it. Both he and the education department officials (none of whom of course have any knowledge of fossils or would be in the least competent to pass on the specimen: that is why it was turned over to them and not the Geological Survey where it should go) had been informed of the date of my arrival, and of my movements, in abundant time. But I'll bet that it will be on its way to Batavia when I get to Sourabaja, and not arrived at Batavia until after I leave. So much for that.

Matthew was quite right in his supposition. In his letter of December 7 from Madiun he indicates that his conjecture was confirmed:

I got word through the American Consul at Soerabaja that the supposed *Pithecanthropus* skull had been sent to the Dep't Education ("S. T. O. V. I. A.") and I have telegraphed for permission to inspect it next Saturday, the 11th, on my way back from Soerabaja in the steamer, which stops at Batavia and stays there most of the day. So unless they make further difficulties we shall see what we shall see. As I wrote to you, I do not much expect to see a real skull and doubt whether it is a fossil at all. But if possible I must carry out my instructions and do my best to examine it. Then I have to cable and prepare a written report to the Museum, and then T. G., I'll be through that business.

The ship stopped at Batavia (or Djakarta) and the specimen, which *was* a fossil, was examined. It proved to be not the skull of *Pithecanthropus*, but rather the epiphysis, or proximal articular end of the humerus (the upper forelimb bone) of a fossil elephant known as a stegodont. Of course the articular surface of the bone was smooth and rounded, something like a skull, but there was no cavity in which a brain might have been lodged. Instead, the "under" side of the bone was a mass of bony tissue. One wonders how anybody trained in anatomy could have made such a colossal mistake.

Will Matthew boarded the ship and sailed on for his meeting with Kate in Egypt. And in spite of his lost time, justifiable annoyance, and disappointment, he was not bitter or angry about the incident. In fact, after he returned home he wrote a ballad.

A Ballad of Pithecanthropus

Souvenir of a visit to Java, December 1926, and of the genial
and generous scientists and other friends whom
it was my privilege to meet there.

Far back in prehistoric time,
   A million years ago
(Or maybe ten, for these learned men
   Don't mind a cipher or so),
A wise old ape came gangling down,
   Swinging from tree to tree,
Till he came to a stop in a high tree-top
   By the shores of the Sunda Sea.

He came from somewhere far to the north,
    Where the mountains were rising high
And the wintry wind was most unkind—
    Cold blasts from a freezing sky.
He had lived up there for a long, long time
    And shivered the winter through,
But as he grew old he felt the cold,
    And concluded it wouldn't do.

"I'll certainly have to go south," he said
    "And escape these winters drear;
And get to a land of tropical sand
    Where it's summer throughout the year.
For I haven't a heavy coat of fur
    To protect me against the weather.
The hair's too thin that covers my skin
    When I pose in the altogether."

So he came to the lovely Sunda shores
    With perpetual summer blest,
Where the balmy breeze and the summer seas
    Invite the trav'ler to rest.
The cocoanuts hung in the stately palms
    All scattered along the shore,
And a varied loot of tropical fruit
    The trees of the forest bore.

Thus he lived an easy and lazy life,
    With plenty of food at hand;
With so much good cheer he grew fatter each year,
    So fat he could hardly stand.
No longer he climbed to the top of the palms
    For his tiffin a nut to cop,
But he lolled at ease at the foot of the trees
    And waited for it to drop.

He made so little use of his brain
    That it wasted slowly away,
And a cellular zone of solid bone
    Took the place of his brain-cells gray,
For 'Nature, the dear old nurse' provides
    For the needs of man and brute,

And he needed a solid skull to withstand
  The shock of the falling fruit.

And so, in the course of time, he died,
  And left his bones in the land,
With his solid head in a lower bed,
  Deep buried beneath the sand.
A million years or so passed by
  While the land was changed, and the sea,
  And the ocean flows where the dry land rose
And the mountains rose from the lea.

Then the buried skull of the ancient ape
  Was brought to the light again
By a slow uplift and a river swift
  That washed it out on the plain.
A native coolie picked it up,
  This fossil of ancient race;
It was sold for ten cents to a group of gents
  Who were visiting 'round the place.

But one of the crowd was a wise old doc,
  And often of men he had known
Whose heads contain, instead of a brain,
  A mass of solid bone.
So he promptly spotted the fossil skull,
  And he said, "This ancient cuss
I really think is the Missing Link,
  The Pithecanthropus."

The word was sent out all over the world.
  And the scientists danced with glee,
And they all expressed their interest
  In the fossil from over the sea.
Said Elliot Smith, "I expect it will prove
  A marvellous thick-skulled and bony 'un."
Said Doctor Hrdlicka, "A few little bits can
  surely be spared the Smithsonian."

Said Osborn, "His brain will undoubtedly show
  The primal beginnings of Art.
The hormones that set this anthropoid pet
  From the lower mammals apart."

"I cannot admit," Doctor Gregory said,
    "That man's Pithecanthropus progeny."
Quoth Sir Arthur Keith, "We'll examine its teeth,
    And learn all about its phylogeny."

Though the lady from Piltdown still held the first place
    And was never quite out of his mind,
Sir Arthur Smith-Woodward had also a good word
    To say for the Heberlein find.
M'sieur Marcellin Boule shrugged his shoulders: "Ah oui,
    Quel dommage qu'on n'a point de squelette!"
While each learned Hun (for they all think as one)
    Said, "Twas foundt py a Cherman, you bet."

The doctor had figured on selling his find
    For a million florins or so,
For a wealthy museum, as soon as they'd see him,
    Could never resist, d'you know.
But a stern official in buttons and tabs
    Held up a warning hand.
"An antiquitee is this fossil," said he,
    "And it must not leave the land."

"You'll have to turn it over," he said,
    To our Office of Education.
We'll study with care this specimen rare
    At the Stovia, for the nation."
Doc Heberlein groaned, as well he might,
    When he saw his visions fair
Of oceans of wealth and unlimited pelf
    All vanishing into the air.

So they brought it in state to Batavia's gate,
    And 'a hundred pipers and all'
Escorted it through the streets on view
    Till they got it within the hall.
They laid it out on a table broad,
    With skulls of every race
Of the anthropoid sort and others that ought
    To compare, to determine its place.

There were calipers large and calipers small
    And reams of paper and pens,

With learned zoologists and anthropologists
    And photographers with lens.
But when they examined the fossil rare,
    They were totally stumped to find,
Where its brain-case swells, a mass of cells
    Of a solid and bony kind.

"Why this," they said, "can never be
    The skull of an ancient ape.
It's too solid a bone, although we own
    It's got 'Pithecanthropus' shape.
So massive a bone as this must come
    From the leg of an elephant old,
A piece of the thigh, or the humeri—
    I fear that we've all been sold."

For they couldn't believe in an anthropoid brain
    Solid bone, even though he were dull;
And Doctor Dubois says the avoirdupois
    Of the fossil's too great for a skull.
And so they concluded that it was not
    The marvellous Missing Link,
But a piece of the joint of a Stegodo (i) nt,—
    And that is what now they think.

If there's any moral to this little tale,
    I think it will have to be:
You mustn't shirk your stint of work
    In a tropical countree.
It's easy to loaf, but if you yield
    Your brain will degenerate,
And solid bone, be it surely known,
    Will fill the whole of your pate.

CHAPTER 20

London Interlude

Will Matthew arrived in Port Said, Egypt, at dinnertime on January 2, 1927, which did not bother Kate whatsoever, even though she was just sitting down with some American friends she had met for the evening repast. She and her friends "took a little boat" and went out to meet the ship. He joined them, they came back to shore for a reunion dinner, and then Kate and Will boarded the P. and O. liner *Macedonia* for a trip along the length of the Mediterranean Sea to Marseilles, and then out through the Strait of Gibralter and up the coast of Europe to London. The journey was uneventful, although at times the seas were rough, and Kate had to endure bouts of seasickness.

They arrived in London on the fourteenth of January, and after having settled themselves in their lodging—the Hotel Ivanhoe, where they were to live for two months—they spent the evening talking about an invitation from the president of the University of California at Berkeley, asking Will to accept a professorship at that institution. It was not a new offer, nor was it a routine affair. The university had been for some time most anxious for Dr. Matthew to come to Berkeley as a replacement for Dr. John C. Merriam, who had left a few years previously to assume the post of president of the Carnegie Institution of Washington—one of the outstanding scientific positions in the United States.

Dr. Merriam first came to Berkeley in the latter years of the nineteenth century, when the university was still relatively small, to study under Joseph LeConte, one of the pioneer naturalists of North America. Thereafter he spent several years in Munich, where he received his doctorate under the great German paleontologist Karl von Zittel. He returned to Berkeley in the 1890s and joined the faculty of the university, and in 1908 he was appointed chairman of the Department of Paleontology. He was an outstanding scholar, a great teacher, an accomplished field man, and a vigorous administrator.

After Merriam went to Washington, his function as a mammalian paleontologist was taken over by his former student Chester Stock, who had obtained his doctorate in 1917 under Merriam's guidance. And in the early 1920s Charles Camp, who had just received his degree under William King Gregory at Columbia University, joined the faculty as a specialist on the so-called lower vertebrates, particularly the fossil reptiles. Along with these changes in personnel, the university in 1920 had merged the Department of Paleontology with the Department of Geology, an organizational arrangement that had existed before 1908.

In 1926 Chester Stock, who since 1921 had been an assistant professor, was lured away to the California Institute of Technology where he was a full professor. To fill the vacancy in mammalian paleontology, the university proposed to bring to the institution a man of mature scientific accomplishment in the field, and to re-create a Department of Paleontology, once again giving it independent status. Will Matthew was the man picked for this position. He had the seniority and the worldwide reputation that, as head of the reconstituted department, would assure the continued preeminence of paleontology at Berkeley. Charles Camp, a young man of brilliant promise, would be a most excellent lieutenant to Matthew, devoting his paleontological talents to the reptiles and other "cold-blooded" vertebrates.

The offer from California now being considered by Will and Kate was no surprise, because feelers had been put out previously. The Matthews had already considered it, and it was tempting. But Matthew loved the house in Hastings-on-Hudson, to which they had given so much time and care and devotion; moreover, Will had spent his entire professional career at the American Museum and was bound by strong feelings of loyalty to the institution and to his friends there. And for the past two years Kate's sister Mame and her elderly mother had been living in a very comfortable and spacious apartment, which Will had modified to suit their particular needs, on the top floor of the Hastings house. Will felt quite strongly that he had a moral responsibility for Kate's mother; indeed, he stated quite un-equivocally that there would be no move from the Hastings house so long as she was with them.

Yet at the Museum Will had a worrisome problem—named Henry Fairfield Osborn. Through the years Osborn had become increasingly domineering, the result in part of his outstanding position in the scientific world, which in turn fed his greater than average bump of vanity, thus making him a difficult colleague. Through the years Matthew had also increased in stature and was no longer a student working under his professor (as perhaps Osborn may still have regarded him) but was a paleonotological giant in his own right. It was hard for Osborn to tolerate this. Indeed, in a letter written in 1921 to his lifelong friend William Berryman Scott of Princeton, Osborn stated "I find my conclusions so different from those of Doctor Matthew, for example, that I could not write a joint work with him. All of his conceptions of evolution are quite different from mine and probably from yours also."[1] It was this situation that finally resulted, as recounted earlier, in Matthew's withdrawal from the planned joint monograph on the evolution of horses.

Thus the opportunity to go west and be quite independent, and to start a paleontological program at the university cut to his own designs, appealed strongly to Will Matthew. And wherever Will would go, Kate would willingly go with him.

On January 22 a cable arrived in London from Kate's brother, Charlie, saying "Mother passed away peacefully last night. Earnest desire of all that you continue trip." It was a shock to Kate and to Will, yet not totally unexpected. A beloved member of the family would no longer be a part of their household. At the same time, the death of Janie Lee removed one of the bonds that held the Matthews to the house in Hastings.

[1] Rainger, *Just Before Simpson,* p. 453.

Within a few days, on January 29, a cable arrived from Professor Osborn. "Congratulations on your call to California—advise return, as soon as Siwalik is finished." So, it would appear, another tie was being cut. This was clear evidence that Osborn would be very satisfied to have Matthew at another institution. It seemed so to Kate, who wrote in her journal "that looks as though the question of our going is almost settled."

Shortly afterward, on February 4, President Campbell in Berkeley sent a cable asking Matthew to meet him in New York as soon as possible, namely, on February 14, 15 or 16. Matthew cabled back, saying that he felt obliged to complete his Siwalik studies in London and would prefer meeting Dr. Campbell in California on March 30. Shortly thereafter, on February 8, Dr. Campbell cabled Matthew, agreeing to meet him in San Francisco at the end of March.

Before the arrival of this second cablegram from President Campbell the Matthews received a momentous letter from their daughter, Elizabeth, announcing her engagement to Ira Nichols, a young medical student completing his training in psychiatry at the College of Physicians and Surgeons of Columbia University. So another tie with Hastings was severed. Events destined to change the course of life for Will and Kate were happening with breathtaking rapidity. For them there seemed to be every reason to accept the invitation from California. The die was cast.

With the decision to make the change from New York to Berkeley essentially settled in their minds, they could now devote their thoughts to their planned activities in London. For Will it meant assiduous attention to the Siwalik mammals at the British Museum; for Kate it meant various expeditions in and around London to see all the things to be seen and visit all the museums and monuments to be visited in this great hub of the British Empire. In spite of the long journey behind them, Kate still retained her unbelievable energy and her undiminished capacity for touring what Will, during an earlier phase of their trip, called "T. and T."—tombs and temples. This harkened back to the time when Kate was dragging Virginia across the breadth of India. She had written to Will mildly complaining of Virginia's waning enthusiasm for going places and seeing things. Will had written back to her, gently admonishing her to have some respect for poor Virginia's inability to keep up with the strenuous schedule that Kate could follow.

In addition to working at the museum and sight-seeing in London, the Matthews found themselves involved in a rather full social life. Numerous scientific people in London wished to entertain the Matthews, and so Will and Kate often attended dinners, went to evening functions, and spent weekends with some of their English friends.

Professor D. M. S. Watson, of the University of London, and his wife were especially attentive, and the Matthews frequently dined with them, usually at their home. Will and Kate were always pleased to receive an invitation from the Watsons; Professor Watson was a vigorous, exciting, charming person, brimming with ideas, scientific and otherwise. The Matthews also enjoyed Professor and Mrs. Elliot Smith, he being one of the leading anthropologists in England at that time. On February 6 at tea with the Elliot Smiths some serious talk focused on the prospect of the Matthews going to Berkeley, and Elliot Smith strongly urged Will to accept the offer. At this stage Will Matthew did not need much urging; he had virtually made up his mind to go, particularly if the university would agree to certain stipulations he was proposing.

Later, on March 12, Matthew would receive a letter from President Campbell agreeing to the terms that Matthew had set forth. The result was that months later Matthew went to Berkeley, where among other things he would organize a new Department of Paleontology, of which he would be the chairman. Paleontology was, and still is, a sort of "in-between" discipline—not geology and not biology, but a fusion of knowledge and techniques from both fields. Matthew was bound for Berkeley to make the study of fossils a discipline in its own right, and he did just that.

In addition to this revival of paleontology, Matthew arranged that, notwithstanding his teaching and administrative duties, he would be afforded ample time for research, including free summers during which he would return to New York to complete long-term projects at the American Museum of Natural History. Evidently these matters were settled to his satisfaction.

Earlier, the Matthews spent the weekend of February 19–20 at Hayward's Heath in Sussex with Sir Arthur and Lady Smith Woodward—Sir Arthur being the dean of British paleontologists and recently retired from the British Museum. They also weekended with the Pilgrims, Guy Pilgrim having recently returned to England from India. Other people with whom they socialized included George Gaylord Simpson, the young man destined to succeed Matthew at the American Museum.

An event that affected Will Matthew profoundly was the receipt of news on February 18 of the death of his beloved Columbia professor and mentor, James Furman Kemp. In the 1890s young Will Matthew had been a particular protégé of Professor Kemp—almost as a son to the older man. For years after Matthew received his doctorate from Columbia, Professor Kemp kept in close touch with the former student who had come to Columbia with the intention of making a career in hard-rock geology but who had changed the direction of his interests, owing to the influence of Osborn, and had gone on to establish a towering reputation in vertebrate paleontology. On the evening of that February day when he learned of Professor Kemp's death, he and Kate attended a dinner of the Geological Society of London. Matthew had to reply to a toast, which he did with a very nice speech, but as Kate noted in her journal, "his mind was on Professor Kemp."

So the days and the evenings in London sped on toward the date of their departure. On March 14 the Watsons gave a farewell dinner at their home. On March 15 they were largely occupied with packing and making the final arrangements for their trip. And on March 16 they boarded the S. S. *Olympic*, bound for New York.

Nothing of particular note happened on their homeward trip; it was similar to most trans-Atlantic voyages of those days—a time of passing the hours pleasantly and anticipating their homecoming. They were looking forward to reunions with the children, from whom they had been separated for many, many months—a whole year for Will. And for Kate there was the sadness of not finding her mother waiting for them, anxious to hear all about their travels.

Above all, they contemplated with pleasure their new life out by San Francisco Bay. There were busy months ahead before they would be settled into that new life.

CHAPTER 21

Museum to University

*T*he Matthew homecoming on March 23, 1927, had bittersweet overtones. Kate and Will felt joy upon returning to the house where they had spent almost all their married life and where the children had spent their formative years, but also sadness in knowing that soon this would no longer be their home—it had to be replaced by a new home, as yet undetermined, thousands of miles away. Dominant, however, was the pleasure of once again being with the children.

They had a happy reunion with Elizabeth, now engaged, when the ship docked at the Hudson River pier. She had spent the winter in New York, working in the library of the American Museum. Margaret and Bill were in Portsmouth, Virginia, with the Charlie Matthew family—Margaret completing her final year of high school, and Bill, his last year of junior high. In short order Will and Kate took the train to Virginia to see their younger children. Will must have been very happy at being reunited with all three of his children after a full year away.

In Hastings and New York City numerous matters required their attention. Moving from their beloved home and severing Will's relationship of more than thirty years with the museum posed problems that at times must have seemed overwhelming. Daunting as these tasks may have been, they were successfully accomplished, so that by mid-June Will had departed for California, and Kate and Elizabeth had arrived in Portsmouth to pick up Margaret and Bill for the transcontinental journey. A great deal of work had to be done before mother and daughter could depart from Hastings. During the last three weeks they closed the sale of the house, had the furniture and other household goods packed and shipped to Berkeley, purchased a Dodge sedan, equipped it with camping gear, and, in Elizabeth's case learned to drive and obtained a driver's license!

The four Matthews left Virginia for a month-long trek across the country. Those were the days before superhighways—indeed, before paved roads on all major routes—and before motels and other roadside conveniences. So they had a long trip, camping out every night, often with the permission of agreeable farmers, from whom they bought fresh milk and eggs for their meals. They prepared *all* their meals, every day. What with making camp and breaking camp, cooking breakfast and dinner, and washing the dishes, it is understandable that their daily mileage was not impressive, especially on the western part of the journey where much of the travel was on dirt or gravel roads.

But they forged on, visiting national parks along their route. They stopped in Nebraska to visit Bill Thomson at Snake Creek. In Oregon the car skidded on a gravel road and went into a ditch, and Kate banged her head on the windshield. She suffered a concussion and was incoherent for several hours, much to the dismay and terror of Betty, Margaret, and Bill. But by the next morning she had recovered and they continued their trip, making the final stretch of the journey along the Pacific coast down to Berkeley. There they were greeted by Will, who had arranged for the whole family to spend a very pleasant few weeks at the Claremont Hotel while they got used to their new environment.

While Kate, Elizabeth, Margaret, and Bill were wending their way across plains and mountains in their modern, gas-propelled prairie schooner, Will Matthew was making a more direct crossing of the continent by train, with stops here and there to visit museums and see fossil collections. Yet even with the delays occasioned by such stops, he reached Berkeley well before the arrival of his family, and this gave him the opportunity to get established in his new quarters at the university. At that time the collection of fossils and all facilities for paleontological and geological activities were located in Bacon Hall, one of the older buildings (long since vanished) on the Berkeley campus.

Bacon Hall—named in honor of Henry Douglas Bacon of Oakland, who donated to the university his "valuable collection of works of art, sculpture and paintings and a library of several thousand volumes" (as he stated in a letter to the board of regents of the university) and also made a gift of twenty-five thousand dollars as partial funding for erection of a building—was completed in 1881. Designed to house the university library and art gallery, it was a rather exuberant Victorian structure of brick and granite, its all-too-fussy front facing west toward the bay, its rounded back toward the Berkeley Hills to the east. That circular rear portion of the building, perforated by very tall, narrow windows, and terminated at the top by a circular, sloping roof crowned with a lantern pierced by numerous small windows, determined to a large degree the interior design of the structure, which had several curved balconies overlooking a central hall.

The geology and paleontology departments had moved into Bacon Hall in 1911 after it had become too crowded and outmoded to serve as a library. The inner space of the building was adapted to uses quite unforeseen by Mr. Bacon and the architects. Dr. Matthew's office was on the second balcony, where perhaps he felt in vaguely familiar but reduced surroundings, since for thirty years in the American Museum he had been ensconced in a huge wing of similar architectural pretensions. During his three years in Berkeley Matthew occupied this office; during the summer of 1930 while he was back in New York the Department of Paleontology was shifted from Bacon Hall to the top floor of the Hearst Mining Building (where the department had its home until 1961, when the new Earth Sciences Building, situated near the north gate of the campus, was completed). The move from Bacon Hall to Hearst Mining was made under the direction of Vertress L. VanderHoof, one of Matthew's graduate students, assisted by Samuel P. Welles, who came to the university in 1930, eventually completing his doctorate under the supervision of Charles Camp.

Having become established in his new quarters at the university, the next task for Will—and for Kate, now that she had arrived in Berkeley with the family—was to find a home. They began their search with the hope of finding a dwelling that would compensate them for the loss of their Hastings house.

At first they were tempted by a charming shingled house of the type known in the architectural world as a "California Brown Shingle," a style of house successfully advocated and frequently constructed by Bernard Maybeck around the turn of the century. This house was just south of the campus and would have been a suitable replacement for the Hastings house, given the period in which it was built and the Edwardian mood it evoked. But Will wanted Kate to have a modern kitchen (in many ways the most important room in a house), and so the shingled house was ruled out.

They settled on a brand new house a few blocks north of the North Gate of the campus, up a steep hill on Cedar Street. It was a Spanish-type house, of stucco with a red tile roof, built in the part of Berkeley that had recently been ravaged by the fire of 1924. It consisted of three wings enclosing a patio, with nicely landscaped grounds that were maintained by a Japanese gardener. For those few years that Will Matthew lived in Berkeley he relied on the gardener for the upkeep of his grounds; gone were the Hastings days, when he spent hours grading, gardening, and trimming. Now he got his exercise by jogging every morning—in which he was about a half-century ahead of his time. The street side of the house consisted of a large upstairs living room above a bedroom and bath for Will and Kate, and a small study. In the middle wing was the dining room and at the corner where the middle and back wings joined was the kitchen. Upstairs in the back wing were two bedrooms and a bath used by Elizabeth and Margaret; below them, the garage and a bedroom that was Bill's domain. It was a convenient and comfortable house and so suited the Matthew family that they had few regrets about leaving Hastings. Their choice of the Cedar Street house, the signing of papers, and the completion of various other legal chores were all accomplished so expeditiously that three weeks after checking into the Claremont Hotel the Matthew family moved into their new home.

For Will Matthew life on Cedar Street and in Bacon Hall, only a few blocks' walk away, was more relaxed than his earlier years in Hastings and at the American Museum of Natural History. For one thing, he no longer labored at grounds maintenance and gardening so that his time at home was more leisurely than previously had been the case. He had loved his work in the garden—it was his form of exercise and relaxation—but in his middle fifties perhaps he liked having a less strenuous physical regimen. Second, he no longer had a two-to three-hour commute as he had each day in New York. Now he walked back and forth, a matter of a few minutes, and he enjoyed the luxury of going home for his midday meal. Third, with a more relaxed daily schedule his evenings were longer and more free than formerly, so that Will and Kate found themselves entertaining at home and visiting others in a manner that had been impossible earlier. Finally, there was a change in Will's workdays. He was busy with teaching and research, and the counseling of students and the management of a department, but he was now on his own, with no higher authority arranging, restricting, or countermanding his plans and activities. In short, he was free from the dominating presence of Henry Fairfield Osborn, and while he did not openly express himself on this matter, almost certainly a weight had been lifted from his shoulders.

Matthew's tenure at the university had financial advantages as well. Not only was he in the highest faculty income bracket, he was also receiving his museum pension. Consequently, he did not feel the financial pressure that had been a part of his life for so many years. This led to a serenity of outlook that was most desirable at his stage of life.

During the first year in Berkeley his two daughters felt that their father should put a bit of his increased income toward the purchase of a new overcoat; the coat he possessed had served him well for seventeen years, it was comfortable, nobody had complained about it, and, to his way of thinking, it was good for many more years to come. But Elizabeth and Margaret were adamant and one day took him in tow and guided him to a haberdashery on Shattuck Avenue in Berkeley. Will Matthew resisted as best he could without making a scene, and when they arrived at the store he even tried to walk past it. But the two young ladies each took an arm and propelled him into the store, where, over his diminishing protests, they directed the selection of a coat and saw to it that he went through with the purchase. In the end he was pleased with the transaction; he could now appear on the streets of Berkeley without looking like an absent-minded, shabby academic. With a new overcoat to protect himself against the chilly winds of a Berkeley evening, Will Matthew was certainly not shabby, but he was very much an academic.

For three decades he had nurtured a desire to teach, a desire that in New York had remained unfulfilled because of his full-time devotion to the paleontological program at the American Museum. There the training of students in the field of vertebrate paleontology had been carried on by Osborn (Matthew being one of his first) and later by W. K. Gregory. Now his chance had come, and he entered into this long-deferred aspect of his professional life with fervor and delight.

From the beginning his efforts were successful and widely appreciated. Osborn had this to say about Matthew's career as a teacher:

> Up to the year 1927, Matthew's achievements were all in the field of exploration and research rather than of education. In this field his training was chiefly that of a writer and speaker with unusual powers of popular exposition and great personal charm because of the original and independent currents of thought lightened by occasional flashes of wit and humor and deepened by his underlying and enthusiastic interest in every subject on which he undertook to speak or to write. . . .
>
> The hopes of Matthew's friends for his success in this new line of scientific endeavor were fully realized. He not only aroused the interest and attention of very large classes of students but he attracted to the none too crowded professional ranks of vertebrate paleontology a number of very promising neophytes.[1]

(Since Osborn's day things have changed; the field of vertebrate paleontology has grown to a size undreamed of by Osborn and his contemporaries; considerable numbers of hopeful young people enter the profession each year.)

Osborn's reference to "very large classes" was inaccurate; it was, in fact, one *very* large class, namely, "Paleontology 1"—a lecture course in general paleontology that during its first year attracted more than seven hundred students to the commodious hall in which it was held. That was a big class for the University of California at Berkeley in those days, and perhaps its size was due in part to an unjustified rumor that a course with such a name would be an educational buggy

---

[1.] Henry Fairfield Osborn, *Bulletin of the Geological Society of America,* 42 (1930):76.

ride. Before the term was out Matthew had disabused his audience of the thought that he was presenting a snap course. Consequently, Paleontology 1 thereafter became known as the "pipe that failed," and subsequent classes were of a more practicable size, attended by students who were truly interested in fossils and what they tell of the history of life.

Matthew wrote a syllabus for the course entitled "Outline and General Principles of the History of Life"—a 253-page paperbound book, published in 1928 and selling for seventy-five cents. Matthew might have written this syllabus as an even-handed review of all life on earth, with a series of more or less equally developed chapters, each devoted to one of the major divisions of animal and plant life. He chose not to do so; his purpose was to introduce students to basic paleontological knowledge and to the general principles of evolution, and then to elaborate on these introductory themes with somewhat detailed discussions of the fossils that were his particular friends, namely, certain Cenozoic mammals. Thus the first fourteen chapters of the syllabus, making up less than one-half its length, are concerned with the definition of fossils, the geologic time scale, the fossil record, the basic classification of animals and plants, evolution, distributions of organisms, and the development of climates. The remaining chapters trace the evolution of life, with one chapter on Cambrian life and the beginning of the fossil record, one chapter on the Paleozoic era and the evolution of land animals, one chapter on dinosaurs and the Age of Reptiles, and one chapter on the Age of Mammals. Following is a series of chapters, each devoted to a group of mammals. One chapter each examines carnivorous mammals, horses, artiodactyls (even-toed hoofed mammals), mastodonts and elephants, extinct mammals of South America, primates—humans and their ancestors and relatives—and the fossil record of humans. The book concludes with a brief chapter discussing the Psychozoic era—the time in which we live—followed by a final chapter titled "The Evolutionary Philosophy."

One may complain that Matthew's presentation was quite lopsided, leaving great areas of the fossil record in limbo. Perhaps so, but the syllabus was an exposition of Matthew's view of evolution, a view based primarily on the history of mammalian evolution. As such, it outlined the knowledge and philosophy of a man who stood as the greatest authority of his time on the evolutionary development of mammals and told an evolutionary story based on an intimate acquaintance with some of the animals involved in that story. The students could go to general textbooks to obtain the record of evolving animals and plants that Matthew did not include in his exposition, but no such text could have the personal appeal provided by Matthew. Through him the student was exposed to interpretations of certain aspects of earth history and evolution, and of certain evolving mammals, from a learned scholar who had spent more than three decades studying these things.

On the last pages of the syllabus Matthew shares a few of his long-considered thoughts with the student reader.

> Evolution in its application to life is a philosophy solidly based upon fact and observation and record. It shows humanity not degenerate from some fancied state of primal innocence, but rising gradually to ever greater heights of achievement, to ever greater perfection of adjustment to the world we live in. . . . The philosophy of evolution looks forward and upward. It is based not upon authority and tradition but upon the sure foundation of fact. It contains no arbitrary commands, only the laws of nature, just but inexorable. It is a cheerful and inspiring view of life to look forward and see the heights yet to be surmounted, to realize that upward progress has been and will be due to the little steps forward achieved by individuals in each generation, and to resolve that each of us will develop the best that is in him, physically, mentally, and socially, as long as he may live, casting aside the imperfections of infancy and youth and fulfilling his destiny in adding something more to the accumulated progress achieved by the countless generations of the past.[2]

In addition to the introductory course in general paleontology Matthew taught two upper-division courses, both limited to vertebrate paleontology. These were "Paleontology 104A—The Osteology, Affinities, and History of the Principal Groups of Mammals" and "Paleontology 104B—History of Vertebrate Life in Western America." Matthew also supervised graduate studies in "Advanced Paleontology."

[2] W. D. Matthew, "Synopsis Lectures in Paleontology—Outline and General Principles of the History of Life," *University of California Syllabus Series,* No. 213 (1928):235–36.

The numbers and names of these courses were holdovers from the Merriam and Stock years at Berkeley, but Matthew revised their content according to his own concepts of what such courses should involve. They were, essentially, courses in mammalian paleontology, the area in which Matthew felt most comfortable. Charles Camp offered a course called "The Osteology and History of the Principle Groups of Fish, Amphibians, and Reptiles," which paralleled Matthew's instruction in mammalian evolution.

Matthew's "promising neophytes" mentioned by Osborn included VanderHoof ("Van"), R. A. Stirton ("Stirt"), Kurt Hesse, Hildegarde Howard, E. Raymond Hall, Remington Kellogg, and William Henry Burt. VanderHoof, Stirton and Howard became active vertebrate paleontologists; Stirton joined the faculty of the University in Berkeley, Van became director of the Santa Barbara Museum, and Hildegarde enjoyed a long tenure at the Los Angeles County Museum, where she became a world authority on the evolution of birds. VanderHoof and Stirton made their reputations in the field of paleomammalogy. Hall, eventually at the University of Kansas, Kellogg, at the United States National Museum, and Burt, at the University of California, were all primarily mammalogists, although some of their research was concerned with fossil forms. Of these several scientists, Van, Stirt and Hildegarde were most closely associated with Matthew. Indeed, Stirton collaborated with Matthew on two of his final scientific papers.

R. A. Stirton, who came to Berkeley from the University of Kansas, as did Kellogg, was throughout his life something of a country boy at heart. He was an indefatigable research scientist and a vigorous field man, and, as fate would have it, he carried on Matthew's program on fossil mammals for many years.

Van, a California native, was more cosmopolitan than Stirt—a quality that served him well in his position as a museum director. He was the student most closely associated socially with the Matthew family and he spent many hours at the Matthew home on Cedar Street.

Hildegarde Howard, who was also very fond of the Matthew family, established a close working relationship with Dr. Loye Miller of the University of California at Los Angeles. She married Henry Wylde, a staff member of the Los Angeles Museum.

Since the first Central Asiatic Expedition to Mongolia in 1922, and up to the time of his decision in 1927 to move to California, Matthew had been very busy studying and describing fossil mammal treasures collected in the Mongolian desert by Walter Granger. The result was the appearance of fifteen papers (thirteen of them between 1923 and 1926), all coauthored with Walter Granger, and two (1928 and 1919) with both Granger and George Gaylord Simpson—Matthew's successor at the American Museum—as coauthors. It was a line of research that Matthew had intended to continue.

There was also the projected study of Siwalik mammals from India, for which the preliminary work had been done in Calcutta and London while Matthew was on his eventful around-the-world trip of 1926 and 1927. His great long-term project, to which he had devoted years of research, a monograph on the Paleocene mammal faunas of the San Juan Basin of New Mexico, was still to be completed and still very much on his mind. Obviously he was, until 1927, contemplating years of research at the American Museum.

Then came the call to Berkeley, and his research plans changed drastically. If he had planned additional contributions on the fossil mammals of Mongolia they were so delayed by his move west that nothing appeared in print. His last Central Asiatic paper was that of 1929 with Granger and Simpson, a description of new mammals of Paleocene age. Perhaps he intended this to be his final Mongolian paper. As for the Siwalik mammals, 1929 saw the publication in the *Bulletin of the American Museum of Natural History* his 124-page paper, "Critical Observations upon Siwalik Mammals." This was a "warm-up" for the large monograph that he never had the opportunity to write. (The study was eventually made by me and was published in 1935.) The San Juan Basin monograph—particularly dear to Will's heart, not just because years of research had been completed, but also because of those many summers in the field—had been provided for in his plans. He had arranged that in ensuing years he would return to New York each summer in order to complete this work. It was not quite finished at the time of his death; I undertook the work of completing it, with advice and direction from Walter Granger and William King Gregory. The publication appeared in 1937 in the large format *Transactions of the American Philosophical Society* (510 pages, 65 plates).

Matthew's research during his three years at Berkeley resulted in a few papers (there was not time for more) dealing with fossil mammals found in California, Nevada, and Texas. In one he described a new fossil hedgehog from Nevada, in another, a mastodont from California, and in a third he discussed the evolution of rhinoceroses—a result of his studies of fossil rhinoceroses from Texas. Two additional papers, both written in collaboration with Stirton, describe, respectively, the rather hyenalike canid, *Borophagus*, and fossil horses from the Pliocene beds of Texas.

In 1924, well before he had any thought of leaving the museum, Matthew had gone to Texas, to collect fossils in the Pliocene Blanco Formation of that state. This was the trip on which George Simpson, having just completed his graduate studies at Yale, was Matthew's assistant and novice driver. In view of the two papers on Blanco fossils that he published at Berkeley, and with the background of his 1924 field trip in that formation, Will Matthew might have gone on to develop a program of fieldwork and research on Texas Pliocene mammals.

In the summer of 1930 Will was in New York, as he had been the two previous summers, doing his final studies on the early mammal faunas from the San Juan Basin of New Mexico. He arrived in July, accompanied by his family, eager to renew his work at the American Museum on the fossils to which he had devoted so many years of study and interpretation. He was installed in his old, familiar office, and it seemed to him, as he settled himself in that room of many memories, that everything had fallen into place for a productive and delightful summer. Among other things, he and Kate were looking forward to the celebration of their twenty-fifth wedding anniversary.

That was not to be, for soon the symptoms of an illness he had first experienced in May recurred. In short, his kidneys were failing, and after an examination by a New York physician, he was advised to return immediately to California.

Accordingly, Will and Kate returned to California by train, while Margaret and Bill went to New Brunswick for a short visit with Christina Matthew, Will's cousin. Then they, too, went home, traveling the width of Canada on the Canadian Pacific, then going south along the coast to Berkeley. Elizabeth, who had married Ira Nichols, was then living in Philadelphia.

Matthew was subjected to a series of grave operations, and one of his kidneys was removed. It was hoped that he would survive this ordeal, but his condition deteriorated, and in his final weeks of life was in great pain.

William Diller Matthew died on September 24, 1930—halfway through his sixtieth year.

CHAPTER 22

Epilogue

D r. Matthew was widely mourned in the scientific community, and numerous expressions of sorrow at his passing and admiration of his significant contributions to his chosen field were published. Especially significant were memorials recorded at the two institutions where he had spent his career.

Immediately after news of his death reached New York the scientific staff of the American Museum of Natural History held a meeting, and the following resolution was passed:

> William Diller Matthew was associated with us during the entire Museum career of nearly all members of the staff, who sadly assemble to honor his memory. Throughout a life of distinguished accomplishment in research and in the upbuilding of the American Museum of Natural History, he was always a ready helper and illuminating counsellor to all who sought him with problems in many fields. Sound, steadfast, patient and brilliant, he was an honor to science and to mankind.
>
> With sorrow and a profound feeling of personal loss, we learn of the death of our friend and colleague. Our heartfelt sympathy goes out to his family and his circle of friends in California.

On September 26 a meeting took place in Bacon Hall where a

group of Matthew's colleagues and friends drew up the following resolution:

> We are meeting here as a small tribute to the memory of our
> friend and colleague, William Diller Matthew. This is the room
> where he worked and studied while a Professor in the University.
> About these walls are his books and specimens—the library which
> he has provided for our use and the exhibits which he has arranged.
> By way of this door, seldom closed, many enjoyed the quiet grace
> of his companionship. From here his inspiring influence has gone
> out, to the Campus and to the World.
>
> The ideals for which he stood will be an inspiration and a
> guide—his achievements, done patiently and without display—his
> scholarship, tempered with dry humor—his keen critical sense,
> balanced by judgment and fine honesty.
>
> As an expression of high regard and in appreciation of his
> friendship, we offer this pledge of our continued interest in the
> aims which he has set so high, and in the plans which he has left
> for us to complete.
>
> (Signed)

Chas. G. Christman	N. E. A. Hinds
R. A. Stirton	Curtis J. Hesse
Bruce L. Clark	F. Earl Turner
A. M. Kinne	William L. Effinger
R. Dana Russell	Andrew C. Lawson
William W. Rand	W. H. Corey
Mary F. Sanborn	Ralph W. Chaney
Helen Nicholson	Barnum Brown
Billie R. Untermann	Chris Jorgensen
G. Leslie Whipple	Chas. L. Camp
S. P. Welles	George D. Louderback
V. L. VanderHoof	

Almost a year after Matthew's friends and colleagues had paid their
separate tributes to him in New York and Berkeley, his ashes were
taken to New Brunswick to be interred in the crypt of the Anglican
church at Gondola Point, on the Kennebecasis River. He was not a
religious man, yet it was fitting that his remains should be returned
to the church he had known as a boy—the church where his parents
were buried. The simple service was attended by Kate and other

members of the family. A pink granite slab, devoid of decorations, was placed in the churchyard there, and on it were inscribed the words

WILLIAM DILLER MATTHEW
HONORED AND BELOVED
BORN FEB. 19 1871
DIED SEPT. 24 1930

The Loyalist scientist, one of the great names in the annals of North American paleontology, had returned to his beloved Canada.

It was August 1931 when this small ceremony took place. Kate had already begun a new life, adjusting to the circumstances that had so changed her view of the years to come. Elizabeth and her husband were living in Philadelphia. Margaret had gone to New York to become a scientific illustrator in the Department of Vertebrate Paleontology at the American Museum.

Kate had stayed in Berkeley until May 1931, when Margaret was graduated from the California School of Arts and Crafts in Oakland, and Bill had finished the semester at high school. Then Kate, after renting the house on Cedar Street, moved to Hartford, Connecticut, to live with her sister, Emma Lee Thayer. Bill was to complete his high school education in Hartford.

Kate and Emma had always been very close, so their arrangement to share a home was eminently sensible; they settled in planning to spend many years together. Emma was an author and artist. She wrote detective novels, quite successfully, and had the added pleasure of designing the dust jackets for her books as they appeared—about two every three years. And while Emma wrote in her study every morning Kate kept the house in order and planned the meals. They had the afternoons and evenings free to do things together. It was a good arrangement.

Then, as they will, things changed.

Ralph Minor was a professor of physics at the University of California, his specialty being light and optics. Because of his particular knowledge in this field he was designated by the university administration to found a school of optometry at Berkeley, which he did most successfully. He was for many years the first dean of the School of Optometry.

Dr. Minor and his wife were close friends of Will and Kate Matthew. In 1930, the year that Will died, Mrs. Minor also passed away. After Kate had moved to Hartford she began to receive letters from Ralph Minor, and in time she was also receiving each week a dozen roses.

They married in California on January 1, 1933, and they moved back into the house on Cedar Street, which had, by prearrangement, been vacated by the tenants.

When they reoccupied the house they insisted that Emma join them, and she did, taking the little room next to the garage, where in previous years young Bill Matthew had lived. Here she resumed her former regimen, devoting each morning to writing detective stories, which continued to flow from her pen until her very old age.

The marriage of Kate to Ralph Minor was a happy union that was to last for twenty-two years, until her death in 1955. Ralph Minor was a gentleman of the old school, a kind and courteous man who was always thoughtful of the people around him. He was loved by his stepchildren, as well as by his own four children, and when he and Kate drove east in the summer, as they did for many years, there were happy reunions at which he was as much a member of the family as the others.

The year after Kate's death, Dr. Minor, Elizabeth, Margaret, Bill, and Gladys, Bill's wife, went to New Brunswick, where a service was held in the church at Gondola Point and where Kate's ashes were interred alongside those of William Diller Matthew. A matching stone was set in the churchyard next to Dr. Matthew's memorial. It carried the following inscription:

KATE LEE

BELOVED WIFE OF	BELOVED WIFE OF
WILLIAM DILLER	RALPH S.
MATTHEW	MINOR
1905–1930	1933–1955
BORN APR. 2 1876	DIED JUNE 20 1955

Ralph Minor lived in the Cedar Street house until his death in 1961. Emma moved into the Women's City Club in Berkeley, and subsequently she lived in Coronado, California, with Ira and Elizabeth Nichols. She died in 1971, about six months short of her one-hundredth birthday.

The work of William Diller Matthew lives on, and today his descriptions and studies of fossil mammals are widely used and respected by a generation of paleontologists more than six decades removed from Matthew. And there is no doubt that Matthew's paleontological works will be treasured and used by future generations of paleontologists.

An appreciation of Matthew has been expressed by one of the present generation of scholars involved with research on mammalian evolution, Dr. Kenneth Rose, of The Johns Hopkins University.

> W. D. Matthew was one of the most influential vertebrate paleontologists of this century. Although he died more than 60 years ago, his work remains vital and inspirational to modern researchers. Attesting to this are widespread references to his publications in even the most recent literature on fossil mammals—particularly that on perissodactyls, carnivores, and early Cenozoic mammals. Matthew is still cited for his contributions on the fossil mammals of Mongolia and the Siwaliks, the West Indies, and South America, but his main focus was North America. To students of early Cenozoic mammals, Matthew's monographs on Paleocene faunas of the San Juan Basin, Wasatchian faunas of the Bighorn and Wind River basins, and Bridgerian carnivores and insectivores are his most important legacy. There he described the fossil mammals from these famous areas thoroughly and meticulously, thereby laying the foundation for all subsequent research on these subjects. But these are much more than descriptive taxonomic reports. With great prescience Matthew also applied his extensive knowledge of anatomy to much broader questions of adaptation and phylogenetic relationships of extinct mammals, and many of his insights have proven valid to this day. Hence these contributions have become indispensable to current workers and are truly classics in the field.

BIBLIOGRAPHY

The following bibliography, edited by Charles Lewis Camp and Vertress Lawrence VanderHoof, is reproduced from William Diller Matthew's *Climate and Evolution,* 2d ed., Special Publications of the New York Academy of Sciences, vol. 1 (New York: The New York Academy of Sciences, 1939). Three additional citations, not recorded in the bibliography, are appended to the end.

ANNOTATED BIBLIOGRAPHY OF
WILLIAM DILLER MATTHEW[1]

By CHARLES LEWIS CAMP AND VERTRESS LAWRENCE VANDERHOOF

1892a. On topaz from Japan. [Columbia College] School of Mines Quarterly **14** (1): 53-56. *4 figs.* 1892.
[Reprinted 1892 in, 'Contributions from the Mineralogical Laboratory of Columbia College **2**.']
[Technical descriptions of crystal faces, cleavages, and angles, of about 100 specimens.]

1893a. On phosphate nodules from the Cambrian of New Brunswick. Trans. N. Y. Acad. Sci. **12**: 108-120. *4 pls.* 10 Ap 1893.
[Reprinted 1893 in, 'Contributions from the Geological Department of Columbia College **9**.']
[Description of thin sections and hypotheses as to origin of the nodules. Sponge spicules and Foraminifera figured.]

1893b. On antennae and other appendages of *Triarthrus beckii*. Am. Jour. Sci. III. **46**: 121-125; and Trans. N. Y. Acad. Sci. **12** (**22**): 237-241. *pl. 7.* 17 Jl 1893.
[Reprinted 1893 in, 'Contributions from the Geological Department of Columbia College **14**.']
[One of the first descriptions of the head appendages of a trilobite. Agrees with Walcott that the group should be classed with the eurypterids and limulids, but "the characters seem to be of a more comprehensive type, approaching the general structure of the other crustacea rather than that of any special form."]

1893c. A study of the scale characters of the northeastern American species of *Cuscuta*. Bull. Torrey Bot. Club **20** (**8**): 310-314. *pls. 164, 165.* 10 Au 1893.
[Reprinted in, 'Contributions from the Herbarium of Columbia College **49**.']
[A key to the species, arranged according to scale characters, is given. Classification of the dodders is still one of the more difficult problems in systematic Botany. Use of scale characters is here for the first time extensively employed.]

1894a. The intrusive rocks near St. John, New Brunswick. Trans. N. Y. Acad. Sci. **13**: 185-203. *4 figs. pl. 5.* 1894.
[Reprinted in, 'Contributions from the Geological Department of Columbia College **22**.']
[Description and areal mapping of certain types of grano-diorite and olivine gabbro.]

1894b. The crystalline rocks near St. John, New Brunswick, Canada. Bull. Nat. Hist. Soc. N. B. **12**: 1-18. *map.* (Read May, 1894.)
"Includes sketch of the principles of metamorphism and igneous intrusions."

1895a. Monazite and orthoclase from South Lyme, Conn. Columbia College School of Mines Quarterly **16** (3): 231-233. Ap 1895.
[Specimens collected during a field trip in spring of 1894.]

1895b. The effusive and dyke rocks near St. John, New Brunswick. Trans. N. Y. Acad. Sci. **14**: 187-217. *2 figs. pls. 12-17.* 28 My 1895.
[Reprinted in, 'Contributions from the Geological Department of Columbia College **30**.']

---

1. Annotations in quotations from William Diller Matthew 1921n. Annotations in brackets by Camp and VanderHoof.

[Petrology, distribution and classification of these types of igneous rocks in the region. Submitted as Thesis for the Degree of Doctor of Philosophy in the Faculty of Pure Science, Columbia College, 1895.]

1895c. The volcanic rocks of the maritime provinces of Canada. Bull. Nat. Hist. Soc., N. B. **13**: 76-83. (Read May 7, 1895.)
[Outlines the volcanic history from Huronian to Triassic and discusses geologic history of the region.]

1896a. Metamorphism of Triassic coals at Egypt, North Carolina. (Abstract) Science **3**: 214. 1896.

1897a. Development of the foot in the *Palæosyopinæ*. Am. Naturalist **31**: 57-58. Ja 1897.
[Matthew's first paper in vertebrate paleontology. Distinguishes between broad and elongate footed types of Eocene titanotheres.]

1897b. Notes on intrusive rocks near St. John, New Brunswick, Canada. Bull. Nat. Hist. Soc., N. B. **15**: 61-64. (Read May 4, 1897.)
[Distribution and analysis of rock samples in continuation of a previous paper on the subject, 1894a.]

1897c. Status of the Puerco fauna. (Abstract) Anat. Anz. **14**: 231-232. 1897. [See also, Science **6**: 852.]

1897d. [Status of the Puerco fauna.] Trans. N. Y. Acad. Sci. **16**: 369-370. 1897.
[Notes on the "rodent" *Mixodectes,* later regarded as an insectivore, and statement that placental mammals become differentiated at beginning of Tertiary.]

1897e. A revision of the Puerco fauna. Bull. Am. Mus. Nat. Hist. **9**: 259-323. *20 figs.* 16 N 1897.
"Revision of the species and their affinities based on a restudy of the Cope collection along with American Museum collections made in 1892 and 1896. Torrejon and Puerco faunas distinguished. Cope's theory of primitive serial carpus and tarsus rejected."

1898a. [*Clænodon* and *Oxyæna* described.] (Abstract) Science **8**: 880. 1898.

1899a. A provisional classification of the freshwater Tertiary of the West. Bull. Am. Mus. Nat. Hist. **12**: 19-75. 31 Mr [-8 Ap] 1899.
"Revised list of mammal faunæ of American Tertiary formations with details of occurrence and critical estimate of the validity and importance of species. Correlation based upon these data."

1899b. Is the White River Tertiary an æolian formation? Am. Naturalist **33 (389)**: 403-408. My 1899.
[Yes, a loessic deposit.]

1899c. **Wortman, Jacob Lawson; & ─────────**
The ancestry of certain members of the Canidæ, the Viverridæ and Procyonidæ. Bull. Am. Mus. Nat. Hist. **12**: 109-139. *10 figs. pl. 6.* 21 Je 1899.
"The sections by Matthew deal with the affinities of the Oligocene ('Miocene') Canidæ of North America and describe *Phlaocyon* as an ancestor of the Raccoons."

1900a. The Cope Pampean collection. Am. Mus. Jour. **1** (2): 24-26. *1 fig.* My 1900.
[This collection formed a part of the Argentine exhibit at the Paris Exposition in 1878. It was purchased by E. D. Cope and later by the American Museum.]

1901a. Additional observations on the Creodonta. Bull. Am. Mus. Nat. Hist. **14**: 1-38. *17 figs.* 31 Ja 1901.
"Classification of the Creodonta, with brief discussion of principles. New observations upon various Creodont genera."
[Favors a compromise between the "horizontal" and "vertical" systems of classification.]

1901b. Fossil mammals of the Tertiary of northeastern Colorado. Mem. Am. Mus. Nat. Hist. **1**: 353-447. *34 figs. pls. 37-39.* N 1901.
"Lacustrine origin of White River and Loup Fork groups rejected in favor of floodplain playa and æolian deposition. New theory of adaptation of sabre-tooth

cats. Description of new collections obtained in 1898, and revision of various groups, Carnivora, camels, oreodonts, etc."

1902a. The hall of fossil vertebrates. Am. Mus. Jour. **2** (**1**) Supplement: 1-19. *11 figs.* Ja 1902.
[See 1903g.]

1902b. A skull of *Dinocyon* from the Miocene of Texas. Bull. Am. Mus. Nat. Hist. **16**: 129-136. *4 figs.* Ap 1902.
[A huge "bear-dog" not greatly resembling either bears or dogs.]

1902c. On the skull of *Bunælurus*, a musteline from the White River Oligocene. Bull. Am. Mus. Nat. Hist. **16** (**12**): 137-140. *3 figs.* 7 Ap 1902.
[A primitive member of the weasel-skunk group, approaching the cats in the reduction of the teeth.]

1902d. [Note on the exhibit, development of the horse.] Am. Mus. Jour. **2** (**5**): 40-41. My 1902. [Unsigned.]

1902e. New Canidæ from the Miocene of Colorado. Bull. Am. Mus. Nat. Hist. **16**: 281-290. *4 figs.* 18 S 1902.
"First outline of theory of Holarctic dispersal and climatic control (286-287)."
[Later elaborated in Matthew "Climate and Evolution" (1915d).]

1902f. A horned rodent from the Colorado Miocene. With a revision of the Mylagauli, beavers, and hares of the American Tertiary. Bull. Am. Mus. Nat. Hist. **16**: 291-310. *17 figs.* 25 S 1902.
"Description of *Ceratogaulus*. Affinities of Mylagaulidæ, *Steneofiber* and *Palæolagus.*"

1902g. The skull of *Hypisodus*, the smallest of the Artiodactyla, with a revision of the Hypertragulidæ. Bull. Am. Mus. Nat. Hist. **16**: 311-316. *4 figs.* 25 S 1902.

1902h. List of the Pleistocene fauna from Hay Springs, Nebraska. Bull. Am. Mus. Nat. Hist. **16**: 317-322. 25 S 1902.

1902i. Review. 'On Vertebrata of the Mid-Cretaceous of the North-West Territory.—(1) Distinctive characters of the Mid-Cretaceous fauna, by Henry Fairfield Osborn.'—(2) 'New genera and species from the Belly River series (Mid-Cretaceous), by Lawrence M. Lambe.' Ottawa Naturalist **16**: 169-170. N 1902.

1903a. Second Cope collection and Pampean collection. Am. Mus. Jour. **3** (**1**): 3-6. Ja 1903. [Unsigned.]
[About 4000 specimens, including a number of Cope's types.]

1903b. The evolution of the horse. Am. Mus. Jour. **3** (**1**) Supplement: 1-30. *4 figs.* 7 *pls.* Guide leaflet **9**. Ja 1903.
[Second edition May 1905; third edition 1909; fourth edition 1913b; fifth edition 1921a.]
"A popular guide to the American Museum collection of fossil Equidæ, pointing out the mechanics of the adaptive changes, and the environmental and climatic changes that brought them about."

1903c. News notes. Am. Mus. Jour. **3** (**4**): 58-61. 1903. [Unsigned.]
[Dept. of Vert. Paleont. 58-61; News notes 62-66.]

1903d. The fauna of the Titanotherium beds at Pipestone Springs, Montana. Bull. Am. Mus. Nat. Hist. **19**: 197-226. *19 figs.* 9 My 1903.
"Various new Lower Oligocene genera described and affinities discussed."

1903e. A fossil hedgehog from the American Oligocene. Bull. Am. Mus. Nat. Hist. **19**: 227-229. *1 fig.* 9 My 1903.
[The first true hedgehog to be found in America.]

1903f. Recent zoopaleontology (concerning the ancestry of the dogs). Science **18** (**440**): 912-913. 5 Je 1903.

1903g. The collection of fossil vertebrates. A guide leaflet to the exhibition halls of vertebrate palæontology. Am. Mus. Jour. **3** (**5**) Supplement: 1-32. *15 pls.* Guide Leaflet **12**. O 1903.
[A revision of 1902a.]

1904a. An extinct cave fauna in Arkansas. Am. Mus. Jour. **4** (1): 6-7. Ja 1904.
[The Conard Fissure.]

1904b. Department of Vertebrate Palæontology: Field explorations in 1903. Am. Mus. Jour. **4** (1): 14-15. 1904. [Unsigned.]

1904c. Eocene fossil mammals of South America. Am. Mus. Jour. **4** (1): 26. 1904. [Unsigned.]
[Casts donated by Florentino Ameghino.]

1904d. A complete skeleton of *Merycodus*. Bull. Am. Mus. Nat. Hist. **20**: 101-129. *21 figs. pl. 3.* 31 Mr 1904.
"Related to *Antilocapra*."

1904e. Notice of two new Oligocene camels. Bull. Am. Mus. Nat. Hist. **20**: 211-215. 2 Je 1904.
[*Pseudolabis* n. gen. and *Miolabis* n. subgen. and key to the genera of fossil camels.]

1904f. —————; & Gidley, James.
New or little known mammals from the Miocene of South Dakota. American Museum Expedition of 1903. II. Carnivora and Rodentia by W. D. Matthew. Bull. Am. Mus. Nat. Hist. **20**: 246-265. *15 figs.* Jl 1904.
[Includes note on evolution of rodent teeth.]

1904g. Outlines of the continents in Tertiary times. Science **19**: 581-582. 1904.
[Also in Am. Geol. **33**: 268-269; treated more fully in Matthew 1906c.]

1904h. Exhibition of a series of foot-bones showing the evolution of the camel. (Abstract) Science **19** (**493**): 892. 10 Je 1904.

1904i. The arboreal ancestry of the Mammalia. Am. Naturalist **38** (**455-456**): 811-819. N-D 1904.
[Huxley and Dollo adduced arboreal ancestry for marsupials. Matthew extends the hypothesis to include placentals.]

1904j. Mode of evolution of the Vertebrata. The New International Encyclopædia. New York, 1904. 'Vertebrata.'
[Also appears in subsequent editions of this work.]

1905a. The fossil carnivores, marsupials, and small mammals in the American Museum of Natural History. Am. Mus. Jour. **5** (1): 22-59. *27 figs.* Ja 1905.
[Reprinted as Guide Leaflet **17**: 1-39.]

1905b. On Eocene Insectivora and on *Pantolestes* in particular. (Abstract) Science **21** (**530**): 298-299. 24 F 1905.

1905c. The mounted skeleton of *Brontosaurus* [in the American Museum of Natural History]. Am. Mus. Jour. **5** (2): 63-70. *4 figs.* Ap 1905.

1905d. Notice of two new genera of mammals from the Oligocene of South Dakota. Bull. Am. Mus. Nat. Hist. **21**: 21-26. *6 figs.* 14 F 1905.
[*Eutypomys* n. gen. of castorid rodents, and *Heteromeryx* n. gen. of Hypertragulidæ; with remarks on distinctions between White River hypertragulids.]

1905e. Notes on the osteology of *Sinopa*, a primitive member of the Hyænodontidæ. Proc. Am. Phil. Soc. **49**: 69-72. 1905.
[A more complete paper on this subject 1906a.]

1905f. [Outlines of the continents in Tertiary times.] (Abstract) Ann. N. Y. Acad. Sci. **16**: 315-316. **1905**.
[See 1906c.]

1906a. The osteology of *Sinopa*, a creodont mammal of the Middle Eocene. Proc. U. S. Nat. Mus. **30**: 203-233. *20 figs. pl. 16.* 1906.
"Description of a complete skeleton in the National Museum; affinities of *Sinopa*, revision of the species."
[A primitive, cursorial creodont related to *Hyaenodon*.]

1906b. Field work of the Department of Vertebrate Palæontology. Am. Mus. Jour. **6** (3): 122-123. Jl 1906. [Unsigned.]

1906c.   Hypothetical outlines of the continents in Tertiary times. Bull. Am. Mus. Nat. Hist. **22**: 353-383. *7 figs.* 25 O 1906.
"Contains a series of maps attempting to correlate views of various authorities on a conservative basis. The Antarctic connections shown on one of these maps were shortly afterwards abandoned as untenable."

1906d.   Fossil Chrysochloridæ in North America. Science **24 (634)**: 786-788. 14 D 1906.
[A humerus from the Rosebud beds, Lower Miocene of South Dakota. Named *Arctoryctes terrenus* n. gen. and sp., without further description. See 1907c: 172.]

1907a.   The skeleton of the Columbian mammoth. Am. Mus. Jour. **7** (1): 5-6. Ja 1907.
[Unsigned and of doubtful authorship.]
[Discovered "four miles east of Jonesboro, Indiana," in a swamp deposit, referred to the Middle Pleistocene.]

1907b.   Department of Vertebrate Palæontology. Field expeditions of 1906. Am. Mus. Jour. **7** (1): 6-8. *1 fig.* Ja 1907. [Unsigned.]

1907c.   A Lower Miocene fauna from South Dakota. Bull. Am. Mus. Nat. Hist. **23**: 169-219. *26 figs.* 14 Mr 1907.
"First discovery of the true Lower Miocene fauna; discussion of correlation and affinities."

1907d.   The *Naosaurus* or "ship-Lizard." Am. Mus. Jour. **7** (3): 36-41. *2 figs.* Mr 1907. [Unsigned.]

1907e.   When brute force ruled the world. Discovery **1** (2): 42. My 1907.

1907f.   The relationship of the 'Sparassodonta.' Geol. Mag. V. **4**: 531-535. D 1907.
[Marsupial relationships favored.]

1908a.   Mammalian migrations between Europe and North America. Am. Jour. Sci. IV. **25**: 68-70. Ja 1908.
[Distribution of Tertiary faunas indicates that Asia "is and has been the great center of evolution and dispersion of dominant mammalian types." Basal Eocene, Middle Eocene and Middle Oligocene indicated as separation between North America and Asia.]

1908b.   *Allosaurus*, a carnivorous dinosaur and its prey. Am. Mus. Jour. **8** (1): 2-5. *1 pl.* Ja 1908.

1908c.   The new *Ichthyosaurus*. Am. Mus. Jour. **8** (1): 6-8. *1 pl.*
[Specimen from Holzmaden showing outline of body.] Ja 1908.

1908d.   A four-horned Pelycosaurian from the Permian of Texas. Bull. Am. Mus. Nat. Hist. **24**: 183-185. *1 fig.* 13 F 1908.
[*Tetraceratops insignis* n. gen. and sp.]

1908e.   Review. 'A revision of the Pelycosauria of North America' by E. C. Case. Science **27 (699)**: 816-818. 22 My 1908.

1908f.   Osteology of *Blastomeryx* and phylogeny of the American Cervidæ. Bull. Am. Mus. Nat. Hist. **24**: 535-562. *15 figs.* 30 Je 1908.
"Description of complete skeleton, discussion of affinities. *Leptomeryx, Blastomeryx, Mazama, Cervus* regarded as successive structural (not genetic) evolutionary stages of the Cervidæ, derived from Palæarctic ancestry and successively invading the New World."

1908g.   News notes. [The Trachodon group] Am. Mus. Jour. **8** (6): 89. O 1908. [Unsigned.]

1908h.   News notes. [Additions to the hall of fossil mammals. Exhibition of *Orohippus osbornianus*.] Am. Mus. Jour. **8** (6): 89. O 1908. [Unsigned.]

1908i.   Fossil fishes. Am. Mus. Jour. **8** (7): 110. N 1908.

1908j.   Exhibit illustrating the evolution of the horse. Am. Mus. Jour. **8** (8): 116-122. *3 figs. 2 pls.* D 1908.

1909a.   Observations upon the genus *Ancodon*. Bull. Am. Mus. Nat. Hist. **26**: 1-7. *5* Ja 1909.

[*Ancodon* (? *Bothriodon*) *leptodus* n. sp. from Lower Rosebud (Arikaree), Lower Miocene of South Dakota.]

1909b. Seventh annual meeting of the American Society of Vertebrate Paleontologists. Science **29** (**735**): 194-198. 29 Ja 1909.
[Signed by Matthew as 'Secretary.' Includes abstracts of papers by Matthew (On a skull of *Apternodus* and the skeleton of a new artiodactyl from the Lower Oligocene of Wyoming); by Matthew and Cook (Pliocene Fauna of Western Nebraska); and by Matthew (On *Bison latifrons*). For the complete paper on *Apternodus* see 1910d.]

1909c. Constitution and By-Laws of the Paleontological Society. 15 Mr 1909.
[Signed by John M. Clarke, T. W. Stanton, David White, C. R. Eastman, W. D. Matthew and H. F. Cleland. Reprinted, without signatures, in Bull. Geol. Soc. Am. **21**: 77-82. 1910. Matthew was elected treasurer to the Society at this time and later became its President.]

1909d. Recent purchases of fossil vertebrates. Am. Mus. Jour. **9** (**3**): 68-69. Mr 1909.
[The Sternberg *Trachodon* "mummy," skeleton of the marine turtle *Toxochelys* and skull of the mosasaur *Clidastes*.]

1909e. The oldest land reptiles of North America. Am. Mus. Jour. **9** (**4**): 91-95. *5 figs.* Ap 1909.
[Notice of the Cope Permian collection.]

1909f. **Osborn, Henry Fairfield; & ⸺**
Geological correlation through vertebrate paleontology by international coöperation. Ann. N. Y. Acad. Sci. **19**: 41-44. 20 Ap 1909.
[First report of Section of Vertebrate Paleontology of International Correlation Committee, National Academy of Sciences. Matthew was the Secretary of this Section. (Abstract) Ann. N. Y. Acad. of Sci. **19**: 302-303.]

1909g. **Osborn, Henry Fairfield; & ⸺**
Explorations and researches of the Department of Vertebrate Palæontology. Preface to: 'Fossil Vertebrates in the American Museum of Natural History, **3**: iii-xvi. 1904-1908.' Ap 1909.

1909h. 'The fossil vertebrates of Belgium' by Louis Dollo [translated by W. D. Matthew]. Ann. N. Y. Acad. Sci. **19**: 99-119. *pls. 4-10.* 31 Jl 1909.

1909i. The Carnivora and Insectivora of the Bridger Basin, Middle Eocene. Mem. Am. Mus. Nat. Hist. **9** (**6**): 289-567. *118 figs. pls. 42-52.* Au 1909.
"Review of the Bridger formation, the source and method of deposition of the sediments, stratigraphic and faunal divisions; adaptation and ecology of the fauna. Detailed osteological descriptions of Bridger Carnivora and Insectivora, discussion of affinities and classification; various new or little known types fully described."

1909j. ⸺; **& Cook, Harold James.**
A Pliocene fauna from western Nebraska. Bull. Am. Mus. Nat. Hist. **26**: 361-414. *27 figs.* 3 S 1909.
"First description of the Snake Creek fauna, regarded as Lower Pliocene."

1909k. Faunal lists of the Tertiary Mammalia of the west. In 'Cenozoic Mammal horizons of Western North America.' By H. F. Osborn. U. S. Geol. Surv. Bull. **361**: 91-120. 1909.
"Lists revised down to 1908, as a basis for a more exact correlation." [The copy in the Matthew Library at the University of California contains Matthew's own manuscript annotations and emendations.]

1909l. Science-History of the universe. Zoology. New York. Current Literature Publishing Co. **6**: 1-194. 1909.
[Contains the following prediction, page 194: "In the present writer's opinion the geological evidence of the ancestry of man and the most direct phylogeny of many mammals will be discovered when the Tertiary formations of Central and Eastern Asia are adequately and thoroughly searched for fossil vertebrates . . ."]

1910a.   Patagonia and the Pampas Cenozoic of South America. A critical review of the correlations of Santiago Roth, 1908. Ann. N. Y. Sci. **19**:149-160. *pl. 14.* 1910.
[Includes list of "Characteristic genera of the South American Cenozoic with provisional ordinal references."]

1910b.   The *Tyrannosaurus*. Am. Mus. Jour. **10** (**1**):2-8. *6 figs.* Ja 1910. [Unsigned.]
[Describes the discovery and collecting of a complete skeleton, later mounted at the American Museum.]

1910c.   A complete pterodactyl skeleton. Am. Mus. Jour. **10** (**2**): 49-50. *1 fig.* F 1910. [Unsigned.]
[*Pterodactylus elegans* from the Solenhofen beds of Bavaria.]

1910d.   On the skull of *Apternodus* and the skeleton of a new artiodactyl. Bull. Am. Mus. Nat. Hist. **28**: 33-42. *5 figs. pl. 6.* 22 Mr 1910.

1910e.   On the osteology and relationships of *Paramys,* and the affinities of the Ischyromyidæ. Bull. Am. Mus. Nat. Hist. **28**: 43-72. *19 figs. pls. 1-5.* 22 Mr 1910.
"Complete skeletons of Eocene rodent *Paramys* and skulls, etc. of other Eocene and Oligocene Rodentia described. The Eocene Ischyromyidæ regarded as the primitive group from which the other rodent groups, all of later age, have been diversely specialized."

1910f.   The pose of sauropodous dinosaurs. Am. Naturalist **44** (**525**): 547-560. S 1910.
[Regards the giant sauropods as wading forms with upright limbs.]

1910g.   The phylogeny of the Felidæ. Bull. Am. Mus. Nat. Hist. **28**: 289-316. *15 figs.* 19 O 1910.
"The Tertiary Felidæ regarded as representing two phyla, one leading from *Dinictis* up into the true cats, the other from *Hoplophoneus* into the sabre-tooth tigers. Nothing is known of their Eocene ancestry. Discussion of the adaptation of the sabre-tooths as shown by characters of skull and skeleton."

1910h.   **Osborn, Henry Fairfield; Gregory, William King; Mosenthel, Johanna Kroeber; & ——————**
Outline classification of the Mammalia recent and extinct. In 'The Age of Mammals.' By H. F. Osborn. New York: Macmillan. 511-563. 1910.
["This classification has been prepared under the direction of the author by William King Gregory and Johanna Kroeber Mosenthal. The Geological range and revision of the extinct genera has been done with the coöperation of William Diller Matthew.—H. F. O."]

1910i.   Notes and literature—Schlosser on Fayûm mammals. Am. Naturalist **44** (**1910**): 700-703. 1910.
"A preliminary notice of Dr. Schlosser's studies upon the collections made in the Oligocene of Egypt for the Stuttgart Museum."
[Abstract in Sci. Progress **5**: 662-663.]

1910j.   Continuity of development. Popular Science Monthly **77**: 473-478. N 1910.
"Contribution to a symposium of the Palæontological Society, dealing with the question of saltation versus continuous progression in phylogeny of fossil vertebrates. The conclusion is reached that the apparently sudden appearance of new types is usually due to migration or to gaps in the geologic record, the best known phyla progressing through very small changes. This is in no wise held to exclude the Mendelian character of such differences, but they are not normally large in amount."

1910k.   The new Plesiosaur. A great marine reptile of the ancient world. Am. Mus. Jour. **10** (**8**):246-250. *2 figs. 2 pls.* D 1910.
[*Cryptoclidus.*]

1911a.   Fort Lee dinosaur. Am. Mus. Jour. **11** (**1**):28-29. Ja 1911.
[Mentions discovery of remains later determined to be of the phytosaur *Clepsysaurus manhattanensis.*]

1911b.   The ground sloth group. Am. Mus. Jour. 11 (4):113-119. *2 figs. 2 pls.* Ap 1911.
          [Figures experimental poses in making the group. Reprinted with additions by the Am. Mus. in 1913.]

1911c.   A tree climbing ruminant. Am. Mus. Jour. 11 (5): 162-163. *1 fig.* My 1911.
          [*Agriochoerus*, an oreodont.]

1911d.   The amphibians of the great coal swamps. Am. Mus. Jour. 11 (6):197-200. *2 figs.* O 1911.

1911e.   Fossil vertebrates—what they teach. Am. Mus. Jour. 11 (7): 246-247. N 1911.
          [Synopsis of three halls of fossil vertebrates in the American Museum.]

1911f.   News notes. [Eocene fossil mammals.] Am. Mus. Jour. 11 (8):311. 1911. [Unsigned.]

1912a.   News notes. [New exhibits of fossil vertebrates.] Am. Mus. Jour. 12 (1):37. Ja 1912. [Unsigned.]

1912b.   News notes. [Prjevalsky wild horses.] Am. Mus. Jour. 12 (2):76. F 1912. [Unsigned.]

1912c.   Review of Abel's 'Grundzüge der Palæobiologie der Wirbelthiere.' Science 35 (896): 341-342. 1 Mr 1912.

1912d.   Florentino Ameghino. Pop. Sci. Monthly 80 (3): 303-307. *3 figs.* Mr 1912.

1912e.   The new four-toed horse skeleton. Am. Mus. Jour. 12 (5): 186. *1 fig.* My 1912.

1912f.   African mammals. Bull. Geol. Soc. Am. 23: 85, 156-162. 1 Je 1912.
          [Discussion of the Fayûm and other fossil faunas. Origin and migrations of mammals.]

1912g.   [Ten years progress in vertebrate paleontology:] Carnivora and Rodentia. Bull. Geol. Soc. Am. 23: 85, 181-187. 1912.
          "Critical reviews of the status of palæontology in these two fields."

1912h.   New dinosaurs for the American Museum. Am. Mus. Jour. 12 (6): 219. O 1912.
          [From the Cretaceous of Red Deer River, Canada.]

1912i.   Facts and theories relating to the ancestry of man. Am. Mus. Jour. 12 (7): 255-256. N 1912.
          [Critical mention of recent work of Schlosser, Boule, Elliot Smith, W. K. Gregory, Hrdlička, G. F. Wright, and others.]

1912j.   News notes. Fossil walrus skull from Penobscot Bay. Am. Mus. Jour. 12 (7): 269. N 1912. [Unsigned.]
          [*Trichechus rosmarus.*]

1912k.   The ancestry of the edentates as illustrated by the skeleton of *Hapalops*, a Tertiary ancestor of the ground sloths. Am. Mus. Jour. 12 (8): 300-303. *2 figs.* D 1912.
          [Reprinted with additions by the American Museum in 1913; pp. 1-8. *4 figs.*]
          [Includes an illustrated phylogeny of the edentates.]

1912l.   [Review of] "The Origin and Antiquity of Man." By G. F. Wright. Current Anthropological Literature 1 (4): 267-272. O-D 1912.

1913a.   News notes. A tiny fossil skull. Am. Mus. Jour. 12 (1): 48. Ja 1913. [Unsigned.]
          [Insectivore *Palæoryctes* from the Torrejon formation. Basal Eocene of New Mexico. See 1913i.]

1913b.   The evolution of the horse in nature. [Revised and extended edition. With S. H. Chubb.] Am. Mus. Guide Leaflet Series 36: 1-63. *38 figs.* Mr 1913.
          "Illustrations mostly new, text revised."
          [1st edition, 1903b; 4th edition, 1921a.]

1913c.   Cuban fossil mammals; preliminary note. Bull. Geol. Soc. Am. 24: 118-119. 24 Mr 1913.
          [Regards land connections unnecessary for introduction of the Cuban mammalian fauna.]

1913d.   News notes. [A visit to Rancho La Brea.] Am. Mus. Jour. 13 (4):200. Ap 1913. [Unsigned.]

1913e.   News notes. Dinosaur explorations in German East Africa. Am. Mus. Jour.
          13 (4): 200. Ap 1913. [Unsigned.]
1913f.   The laws of nomenclature in paleontology. Science 37 (960): 788-792. 23 My
          1913.
          [Discusses the "special difficulties that beset the vertebrate paleontologist." See
          1913h.]
1913g.   Certain theoretical considerations affecting phylogeny and correlation. Bull. Geol.
          Soc. Am. 24: 283-292. 1 fig. 1 Je 1913.
          [Abstract, page 118.]
          [Theories of dispersal, hybridization during migration and effects of remoteness
          from dispersal centers on diversity of type. Draws parallels from dispersal of
          human races.]
1913h.   Nomenclature in paleontology. Science 38 (968): 87-88. 18 Jl 1913.
          [Continuation of arguments presented in 1913f.]
1913i.   A zalambdodont insectivore from the Basal Eocene. Bull. Am. Mus. Nat. Hist.
          32: 307-314. 6 figs. pls. 60, 61. 25 Jl 1913.
          [Abstract in Ann. N. Y. Acad. Sci. 23: 263-264. Palæoryctes puercensis n. gen. &
          sp.]
1913j.   American Museum expeditions for fossil vertebrates. Am. Mus. Jour. 13 (6):
          286-287. O 1913.
          [Cretaceous dinosaurs, Puerco mammals and work at Agate Springs, Nebraska.]
1913k.   The asphalt group of fossil skeletons. The tar-pits of Rancho La Brea, Cali-
          fornia. Am. Mus. Jour. 13 (7): 290-297. 6 figs. N 1913.
1913l.   News notes. [Exhibit of fossil rhinoceroses.] Am. Mus. Jour. 13 (8): 374. 1913.
          [Unsigned.]
1914a.   News notes. [Fossils from Rancho La Brea.] Am. Mus. Jour. 14 (1): 46-47.
          Ja 1914. [Unsigned.]
1914b.   News notes. [New South American Fossils.] Am. Mus. Jour. 14 (1): 47-48.
          1914. [Unsigned.]
1914c.   Extinct vertebrates. [Report of the] Department of Vertebrate Palæontology.
          45th Annual Report Am. Mus. Nat. Hist. 1913: 57-59. 1914.
1914d.   [Note on Felis atrox bebbi of Rancho La Brea.] Nature 92: 640. 5 F 1914.
          ["Appears to be as nearly related to the tiger as to the lion."]
1914e.   Origin of Argentine wild horses. Nature 92: 661. 12 F 1914.
1914f.   Discussion [of C. Stock's paper on "The systematic position of the mylodont sloths
          from Rancho La Brea."] Bull. Geol. Soc. Am. 25: 144. 30 Mr 1914.
1914g.   Report of progress in the revision of the Lower Eocene faunas. (Abstract) Bull.
          Geol. Soc. Am. 25: 144-145. 30 Mr 1914.
1914h.   Introductory note to Albert Johannsen "Petrographic Analysis of the Bridger,
          Washakie, and other Eocene formations of the Rocky Mountains." Bull. Am.
          Mus. Nat. Hist. 33: 209-210. 31 Mr 1914.
1914i.   News notes. [Two new fossil mammals of the Lower Eocene.] Am. Mus. Jour.
          14 (4): 167-168. Ap 1914. [Unsigned.] [See 1915n.]
1914j.   Notes on Cuban fossil mammals. (Abstract) Ann. N. Y. Acad. Sci. 23: 263-264.
          31 Ap 1914.
1914k.   News notes. [Expeditions for fossil vertebrates.] Am. Mus. Jour. 14 (5): 214.
          My 1914. [Unsigned.]
1914l.   [Notes on auditory ossicles of rodents.] In "The auditory ossicles of American
          rodents." By T. D. A. Cockerell, and others. Bull. Am. Mus. Nat. Hist. 33:
          350-351, 379-380. 14 Jl 1914.
1914m.   Time ratios in the evolution of mammalian phyla. A contribution to the problem
          of the age of the earth. Science 40 (1024): 232-235. 14 Au 1914.
          [Assumption of a "fairly constant maximum rate of progressive evolution" would
          give a yard-stick for measurement of geologic time.]

1914n.   Evidence of the Paleocene vertebrate fauna on the Cretaceous-Tertiary problem. Bull. Geol. Soc. Am. **25**: 381-402. *3 figs.* 15 S 1914.

"Characters of the Paleocene faunæ (Puerco, Torrejon, etc.), comparison with Cretaceous faunæ and with true Eocene faunæ, correlation and interpretation. Faunal migrations are due to widespread diastrophic movements accompanied by climatic changes, and constitute the best practical evidence of such widespread movements. The most marked faunal break is between the Paleocene and Eocene; between the Lance and Belly River there is a very considerable time interval, but no such marked break as should indicate a change in period. Between Lance and Paleocene there is an apparent faunal break but it is of doubtful interpretation. But until these doubtful points are cleared up it seems better to retain the division at this point which is the classical and generally accepted line between Cretaceous and Tertiary."

[See 1921d, 1921i.]

1914o.   News notes. [Discovery of *Moropus* skeletons.] Am. Mus. Jour. **14** (6-7): 269. O-N 1914. [Unsigned.]

1914p.   The largest known dinosaur. A huge extinct reptile from German East Africa, the largest known quadruped. Scientific American **111** (22): 443, 446-447. *1 fig.* 28 N 1914.

[*Gigantosaurus*.]

1914q.   Note on *Felis atrox*. Nature **92**: 640. 1914.

1914r.   New discoveries in the American Eocene. Report of the 83rd meeting, Brit. Assoc. Adv. Sci. 1913: 491. 1914.

[Differentiates the zones and faunas. Speaks of advances in studies on phylogenies and relationships of Tertiary Mammalia.]

1915a.   The Tertiary sedimentary record and its problems. In "Problems of American Geology." By Charles Schuchert, and others. Yale University Press. Chapter 7: 377-478. *40 figs. map.* Ja 1915.

"Discussion of the Tertiary sedimentary formations of the Cordilleran region, their character and origin. Review of fossil record as contained therein showing the evolution of the Tertiary faunas."

1915b.   News notes. [New exhibits of fossil vertebrates.] Am. Mus. Jour. **15** (1): 32. Ja 1915. [Unsigned.]

1915c.   News notes. [New exhibits of fossil mammals.] Am. Mus. Jour. **15** (2): 86. F 1915. [Unsigned.]

1915d.   Climate and Evolution. Ann. N. Y. Acad. Sci. **24**: 171-318. *33 figs.* 18 F 1915. [The author's own copy of this famous work, in the Matthew Library, is extensively annotated.]

1915e.   —————; & **Granger, Walter.**

A revision of the Lower Eocene Wasatch and Wind River faunas. Part I. Order Feræ (Carnivora) Suborder Creodonta. By W. D. Matthew. Bull. Am. Mus. Nat. Hist. **34**: 1-103. *87 figs.* 3 Mr 1915.

[For continuation see, Matthew 1915l, 1915n, and 1918k.]

1915f.   **de la Torre, Carlos; &** —————

Megalocnus and other Cuban ground sloths. (Abstract) Bull. Geol. Soc. Am. **26**: 152. 31 Mr 1915.

[*Megalocnus, Mesocnus, Miocnus,* and *Microcnus* from the Cuban Pleistocene. Nearest continental ally is *Megalonyx*. See 1931c.]

1915g.   Affinities of *Hyopsodus*. (Abstract) Bull. Geol. Soc. Am. **26**: 152. 31 Mr 1915. [Referred to the Condylarthra.]

1915h.   Reconstruction of the skeleton of *Brachiosaurus*. (Abstract) Bull. Geol. Soc. Am. **20**: 153. 31 Mr 1915.

[From German East Africa. The largest known dinosaur.]

1915i.   Ground-sloth from a cave in Patagonia. Am. Mus. Jour. **15** (5): 256. My 1915. [*Grypotherium* as a contemporary of man.]

1915j. News notes. [Richard Lydekker, obituary notice.] Am. Mus. Jour. **15** (**5**):263. My 1915. [Unsigned.]

1915k. Some remarkable extinct animals of South America. (Abstract) Ann. N. Y. Acad. Sci. **24**:355. 14 My 1915.

1915l. ————; & Granger, Walter.
A revision of the Lower Eocene Wasatch and Wind River Faunas. Part II. Order Condylarthra, Family Hyopsodontidæ. By W. D. Matthew. Bull. Am. Mus. Nat. Hist. **34**:311-328. *10 figs.* 29 My 1915.

1915m. Dinosaurs. With special reference to the American Museum collections. Am. Mus. Handbook Series **5**:1-124; 160-162. *48 figs. 1 pl.*
"A semi-popular review of the various kinds of dinosaurs, their antiquity, habits and environment, based chiefly upon American Museum collections."
[Contains chapters by S. W. Williston, H. F. Osborn and Barnum Brown.]

1915n. ————; & Granger, Walter.
A revision of the Lower Eocene Wasatch and Wind River faunas. Part IV. Entelonychia, Primates, Insectivora (Part). Bull. Am. Mus. Nat. Hist. **34**:429-483. *52 figs. pl. 15.* 24 S 1915.
"This revision is based upon the extensive collections and exact stratigraphic work of the American Museum expeditions of 1905, 1909-1916. Numerous new or little known mammals are described, and the affinities of all the Lower Eocene genera and species discussed; the phyletic and faunal changes and the successive levels in the Lower Eocene formations and their correlation fully shown. Part III by Granger covers the remaining Condylarthra; Part V, is listed under the year 1918 [1918]; Part VI, Tillodonta and Tæniodonta by Granger; Part VII, amblypods by Granger; Part VIII, Perissodactyla by Granger, and Part IX, conclusions, tabulation and correlation of faunas by Matthew and Granger not yet published."

1915o. News notes. [The Fulton mastodon.] Am. Mus. Jour. **15** (**7**):373. N 1915. [Unsigned.]

1915p. Mammoths and mastodons. A guide to the collection of fossil proboscideans in the American Museum of Natural History. Am. Mus. Guide Leaflet Series **43**:1-26. *11 figs. 1 pl.* N 1915.
[Author's annotated copy in the Matthew Library.]

1915q. ————; & Barnum Brown.
Corythosaurus, the new duck-billed dinosaur. Am. Mus. Jour. **15** (**8**):427-428. D 1915.

1915r. News notes. [Fossil Sirenian from Porto Rico.] Am. Mus. Jour. **15** (**8**):432. D 1915. [Unsigned.]

1915s. Cope, Edward Drinker; & ————
Hitherto unpublished plates of Tertiary Mammalia and Permian Vertebrata. Am. Mus. Nat. Hist. Monogr. Ser. **2**: Unpaged series of plates, with explanations. 1915.
[Published by the American Museum from plates in the custody of the U. S. Geol. Survey. The plates were prepared under the direction of E. D. Cope for the U. S. G. S. The descriptions of plates were written by Matthew.]

1915t. New discoveries in the Lower Eocene mammals. (Abstract) Ann. N. Y. Acad. Sci. **24**:383. 14 My 1915.

1915u. ————; & Clarke, John M.
Peccaries of the Pleistocene of New York. Bull. Geol. Soc. Am. **26**:150-151. 31 Mr 1915.

1915v. [General consideration of paleontologic criteria in determining time relations.] (Discussion.) Bull. Geol. Soc. Am. **26**:411. 23 N 1915.
[Problem of correlation by use of vertebrates. By title.]

1915w. [Report on fossil vertebrates of eastern Texas.] In "Problem of the Texas Tertiary sands." By E. T. Dumble. Bull. Geol. Soc. Am. **26**:470-472. 4 D 1915.
[See 1920m.]

1915x.   **Osborn, Henry Fairfield; &** ——————
Exploration and researches of the Department of Vertebrate Palæontology. Preface
to "Fossil vertebrates in the American Museum of Natural History." **4**: iii-xvi.
1909-1912. Ap 1915.

1916a.   **Eastman, Charles R.; William King Gregory; &** ——————
Record of progress in vertebrate palæontology. Science **43** (**1099**): 103-110. 21
Ja 1916.

1916b.   In "Some remarks on Matthew's 'Climate and Evolution,' with Supplemental Note
by W. D. Matthew." By Thomas Barbour. Ann. N. Y. Acad. Sci. **27**: 11-15.
26 Ja 1916.
[Discusses the origin of insular mammals.]

1916c.   A new sirenian from the Tertiary of Porto Rico, West Indies. Ann. N. Y. Acad.
Sci. **27**: 23-29. *2 figs.* 28 Ja 1916.
[Abstract, *ibid.* **26**: 439. 1915.]
[? *Halitherium antillense*, n. sp.]

1916d.   The horse and his progenitors. Science Conspectus [Boston] **6** (**1**): 1-15. *9 figs.
1 pl.*

1916e.   The grim wolf of the tar-pits. Am. Mus. Jour. **16** (**1**): 45-47. *2 figs.* Ja 1916.
"The great extinct wolf from the asphalt deposits at Rancho La Brea near Los
Angeles."

1916f.   News notes. [Dinosaurs from Alberta.] Am. Mus. Jour. **16** (**1**): 74-75. Ja 1916.

1916g.   News notes. [Fossil reptile exhibits; *Myotragus*, Antillean fossils.] Am. Mus.
Jour. **16** (**2**): 139. F 1916. [Unsigned.]

1916h.   News notes. [Charles Falkenbach, obituary notice.] Am. Mus. Jour. **16** (**3**):
209. Mr 1916. [Unsigned.]

1916i.   News notes. [Expeditions for fossil vertebrates.] Am. Mus. Jour. **16** (**3**): 209-
211. Mr 1916. [Unsigned.]

1916j.   A reptilian aëronaut. A new skeleton of *Pteranodon*, the giant flying reptile of the
Cretaceous period. Am. Mus. Jour. **16** (**4**): 251-252. *1 fig.* Ap 1916.

1916k.   The origin of the Pacific island faunas. Science **43** (**1115**): 686. 12 My 1916.

1916l.   A marsupial from the Belly River Cretaceous. With critical observations upon the
affinities of the Cretaceous mammals. Bull. Am. Mus. Nat. Hist. **35**: 477-500.
4 figs. pls. 2-6. 24 Jl 1916.
"Description of a new genus *Eodelphis*, related to the opossums, redescription of
Thlaeodon padanicus Cope and critical estimate of the remaining Cretaceous
mammals with remarks upon the interpretation of the marsupial dentition."

1916m.   Methods of correlation by fossil vertebrates. Bull. Geol. Soc. Am. **27**: 515-524.
1 S 1916.
"Discussion of principles of correlation. The Equidæ are regarded as affording the
best standard for Tertiary correlation."

1916n.   News notes. [Expeditions for fossil vertebrates.] Am. Mus. Jour. **16** (**6**): 412.
O 1916. [Unsigned.]

1916o.   Scourge of the Santa Monica Mountains [California]. Am. Mus. Jour. **16** (**7**):
469-472. *2 figs.* N 1916.
[The sabre-tooth cat, *Smilodon*.]

1916p.   Kunz on Ivory and the Elephant. Am. Mus. Jour. **16** (**8**): 485-495. *8 pls.*

1916q.   **Osborn, Henry Fairfield; &** ——————
Explorations and researches of the Department of Vertebrate Palæontology. Preface to "Fossil Vertebrates in the American Museum of Natural History **5**: iii-viii.
1913-1914." 1916.

1917a.   News notes. [Fossil tree trunks from the Cretaceous of Alberta.] Am. Mus.
Jour. **17** (**1**): 78. Ja 1917. [Unsigned.]

1917b.   News notes. [A Pliocene mastodon skeleton from Texas.] Am. Mus. Jour. **17**
(**2**): 149. F 1917. [Unsigned.]

1917c. **Eastman, Charles R.; William King Gregory; &** —————
Recent progress in paleontology. Science **45** (**1153**): 117-121. 2 F 1917.

1917d. A fossil deer from Argentina, with a discussion of the distribution of various types of deer in North and South America. Am. Mus. Jour. **17** (**3**): 207-211. *2 figs.* Mr 1917.
"Notice of mounted skeleton of *Brachyceros pampæus*, with discussion of its affinities and the origin of the South American deer."

1917e. Gigantic *Megatherium* from Florida. (Abstract) Bull. Geol. Soc. Am. **28**: 212. 31 Mr 1917.
[This specimen "must have equaled or exceeded in bulk any known land mammal, living or extinct,"—with exception of *Baluchitherium*, which was not known at this date. Complete paper in Matthew, 1917, M. S.]

1917f. —————; **& Granger, Walter.**
The skeleton of *Diatryma*, a gigantic bird from the Lower Eocene of Wyoming. Bull. Am. Mus. Nat. Hist. **37**: 307-326. *1 fig. pls. 20-33.* 28 My 1917.
[Abstracts in Bull. Geol. Soc. Am. **28**: 212; and Science N.S. **46**: 246.]
"Description of a nearly complete skeleton of a gigantic flightless bird from the Wasatch formation of the Bighorn Basin. It has a huge skull and enormous compressed beak like the extinct South American *Phororhachos*, but it is not related to that genus, nor nearly related to any other known type. Its affinities are difficult to estimate; although having some features in common with the ostriches and cassowaries it does not appear to be related to them, but to be a very ancient terrestrial adaptation derived from primitive neognathine ('carinate') stock. Suggestions as to its bearing upon the general problem of the evolutionary history of birds."

1917g. Prehistoric animal life. The Mentor, Dept. Nat. Hist. **5** (**13**): 1-23. *14 figs. 6 pls.* 15 Au 1917.

1917h. A Paleocene bat. Bull. Am. Mus. Nat. Hist. **37**: 569-571. *1 fig.* 7 S 1917.
[*Zanycteris paleocenus* n. gen. & sp. from base of the Wasatch, near Ignacio, Colorado.]

1917i. Absence of the pollex in Perissodactyla. Bull. Am. Mus. Nat. Hist. **37**: 573-577. 7 S 1917.

1917j. Man and the Anthropoid. Science **46** (**1184**): 239-240. 7 S 1917.

1917k. The dentition of *Nothodectes*. Bull. Am. Mus. Nat. Hist. **37**: 831-839. *pls. 99-102.* 5 D 1917.
"*Nothodectes* comes from a new fossil horizon at the top of the Paleocene, and is very closely allied to *Plesiadapis* of the Cernaysian of France. It is related both to Primates and Insectivora, especially to the Menotyphla."

1917l. A giant Eocene bird. Am. Mus. Jour. **17** (**6**): 417-418. *2 figs.* [See 1917f.]

1917m. News notes. [Duck-billed dinosaurs—Census of fossil skeletons.] Am. Mus. Jour. **17** (**6**): 419. O 1917. [Unsigned.]

1918a. **Gregory, William King; &** —————
Vertebrate palæontology. American Year Book. **1917**: 634-636. 1918.

1918b. The mounted skeleton of *Moropus* in the American Museum. A "clawed ungulate" from the middle Tertiary of Nebraska. Am. Mus. Jour. **18** (**2**): 120-123. *1 fig. 1 pl.* F 1918.

1918c. Generic nomenclature of the Proboscidea. (Abstract) Bull. Geol. Soc. Am. **29**: 141. 31 Mr 1918.

1918d. Affinities and phylogeny of the extinct Camelidæ. (Abstract) Bull. Geol. Soc. Am. **29**: 144. 31 M 1918.

1918e. —————; **& Granger, Walter.**
Fossil mammals of the Tiffany Beds. (Abstract) Bull. Geol. Soc. Am. **29**: 152. 31 Mr 1918.
[Base of the Wasatch series in southern Colorado.]

1918f.   Notes on the American Pliocene rhinoceroses. (Abstract) Bull. Geol. Soc. Am. **29**: 153. 31 Mr 1918.

1918g.   Contributions to the Snake Creek fauna, with notes upon the Pleistocene of Western Nebraska, American Museum Expedition of 1916. Bull. Am. Mus. Nat. Hist. **38**: 183-229. *20 figs. pls. 4-10.* 18 Ap 1918.
"Fauna of the Snake Creek pockets recognized as including two distinct faunæ, Upper Miocene and Lower Pliocene, respectively. Various new types described. Revision of rhinoceros genera."

1918h.   Skeletons of the Cuban ground sloth in the Havana and American Museums. Am. Mus. Jour. **18** (4): 312-313. *1 pl.* Ap 1918.
[*Megalocnus.* See 1931c.]

1918i.   A fortunate collector. [William Stein, the discoverer of the giant bird, *Diatryma,* which was found in the Bighorn Basin of Wyoming.] Am. Mus. Jour. **18** (5): 389. *3 figs.* My 1918.

1918j.   A Tertiary alligator. Am. Mus. Jour. **18** (6): 505-506. *1 fig.* O 1918.

1918k.   ————; & **Granger, Walter.**
A revision of the Lower Eocene Wasatch and Wind River Faunas. Part V. Insectivora (continued), Glires, Edentata. By W. D. Matthew. Bull. Am. Mus. Nat. Hist. **38**: 565-657. *68 figs.* 21 D 1918.
"Revision of Insectivora and Glires with description of various new or little known types. Description of a primitive genus of Metacheiromyidæ with discussion of its affinities. Regarded as a very primitive edentate related to early Insectivora and collateral ancestor of armadillos and other Xenarthra on one side, of Pholidota on the other side. Tubulidentata not related to true edentates." [See 1915e.]

1918l.   Affinities and origin of the Antillean mammals. Bull. Geol. Soc. Am. **29**: 657-666. [See 1919g and 1931c.]

1918m.   News notes. [Fossils from Snake Creek beds.] Am. Mus. Jour. **18** (8): 731. D 1918. [Unsigned.]

1918n.   **Osborn, Henry Fairfield; &** ————
Explorations and researches of the Department of Vertebrate Palæontology. Preface to "Fossil Vertebrates in the American Museum of Natural History, 6: iii-vii. 1915-1917." 1918.

1919a.   ————; & **Gregory, William King.**
Vertebrate palæontology. American Year Book, **1918**: 695-696. 1919.

1919b.   Notes. [Lawrence M. Lambe, obituary notice.] Natural History **19** (3): 351-352. Mr 1919.

1919c.   Reply to a letter from John Burroughs. Natural History **19** (4-5): 491-493. Ap-My 1919.

1919d.   Honor to Adam Hermann. Natural History **19** (4-5): 491-493. Ap-My 1919.
"Address on occasion of the retirement of Mr. Hermann, head preparator in the American Museum department of vertebrate palæontology."

1919e.   Notes. [The American Museum expedition to Jamaica.] Natural History **19** (6): 755. D 1919. [Unsigned.]
"Notice of the expedition of November, 1919, pointing out the evidence that the discovery of a fossil fauna will supply on palæogeography."

1919f.   Notes. [Notice of expedition to the Agate fossil quarry.] Natural History **19** (6): 755. D 1919. [Unsigned.]

1919g.   Recent discoveries of fossil vertebrates in the West Indies and their bearing on the origin of the Antillean fauna. Proc. Am. Phil. Soc. **58** (3): 161-181.
[Abstract in Science **49**: 546-547. See 1918l and 1931c.]

1920a.   Plato's Atlantis in palæogeography. Proc. Nat. Acad. Sci. **6** (1): 17-18. Ja 1920.
[No scientific evidence for or against the existence of Atlantis.]

1920b.   Flying reptiles. Natural History **20** (1): 73-81. *3 pl.* Ja-F 1920.

"Notice of the *Pteranodon* skeleton placed on exhibition, with some general remarks on pterodactyls."

1920c. —————; & Clarke, John M.
Supposed fossil horse from the late Pleistocene found at Monroe, Orange Co., New York. (Abstract) Bull. Geol. Soc. Am. **31**: 204. Mr 1920.
[A native *Equus* survived the Wisconsin glaciation in eastern N. A.]

1920d. Status and limits of the Paleocene. (Abstract) Bull. Geol. Soc. Am. **31**: 221. 31 Mr 1920.
[Paleocene "includes the Puerco, Torrejon, Fort Union, and probably . . . the Cernaysian."]

1920e. New specimen of the Pleistocene bear *Arctotherium* from Texas. (Abstract) Bull. Geol. Soc. Am. **31**: 224-225. 31 Mr 1920.

1920f. John Campbell Merriam: New President of the Carnegie Institution. Natural History **20** (3): 253-254. *portrait*. My-Je 1920.

1920g. [Review.] "Die Stämme der Wirbeltiere." By Othenio Abel. Science **52** (**1332**): 37-38. 9 Jl 1920.

1920h. A new genus of rodents from the Middle Eocene. Jour. Mam. **1** (4): 168-169. Au 1920.
[*Reithroparamys* n. gen., type *Paramys delicatissimus* Leidy, from the Bridger.]

1920i. Social evolution: a palæontologist's viewpoint. Natural History **20** (4): 374-378. S-O 1920.
[A discussion of advances in social organization and their bearings on individual conduct.]

1920j. Three-toed horses. Natural History **20** (4): 473-478. *1 fig.* S-O 1920.
[Essay inspired by H. F. Osborn's monograph—"The Oligocene, Miocene, and Pliocene Equidae of North America."]

1920k. Canadian dinosaurs. Natural History **20** (5): 536-544. *3 figs. 3 pls.* N-D 1920.
[*Ankylosaurus, Monoclonius, Deinodon, Corythosaurus, Struthiomimus, Saurolophus, Stephanosaurus, Prosaurolophus, Procheneosaurus, Kritosaurus, Trachodon, Gorgosaurus.*]

1920l. The proofs of the evolution of Man. Natural History **20** (5): 574-575. N-D 1920.
"Scientific evidence would indicate that the human race came from small tree-dwelling ancestors in Central Asia."

1920m. [Report on East Texas vertebrates.] Bull. Univ. Texas **1869**: 225, 231-233.

1921a. Evolution of the horse in nature. Am. Mus. Nat. Hist. Guide Leaflet Series **36** (1): 1-37. *25 figs. 1 pl.* 4th ed. revised. Mr 1920.
[1st edition, 1903b.]

1921b. Notes on scientific museums of Europe. Natural History **21** (2): 185-190. Mr-Ap 1921.

1921c. A note on the Cernaysian mammal fauna. Amer. Jour. Sci. V. **1**: 509-511. Je 1921.

1921d. The Cannonball Lance formation. Science **54**: 27-29. Jl 1921.
[Vertebrate evidence summarized. (See 1914m, 1921i and 1922a.) On this basis the Lance is Cretaceous. By compromise with paleobotanists and stratigraphic geologists the Fort Union might be placed with it in the Cretaceous. The uppermost Paleocene faunas should be regarded as Tertiary.]

1921e. Notes. [Dr. Mook's researches on Crocodilia.] Natural History **21** (4): 433. Jl-Au 1921. [Unsigned.]

1921f. —————; & Granger, Walter.
New genera of Paleocene mammals. Am. Mus. Novitates **13**: 1-7. 6 S 1921.
[*Ectypodus musculus* n. gen. & sp., *Eucosmodon* n. gen., *Paradectes elegans* n. gen. & sp., *Thylacodon pusillus* n. gen. & sp., *Leptacodon tener* n. gen. & sp., *Xenacodon mutilatus* n. gen. & sp., *Acmeodon secans* n. gen. & sp., *Labidolemur*

soricoides n. gen. & sp., *Ignacius frugivorus* n. gen. & sp., *Navajovius kohlhaasæ* n. gen. & sp., *Carpodaptes aulacodon* n. gen. & sp., *Eoconodon* n. gen., *Mixoclænus encinensis* n. gen. & sp.]

1921g. *Stehlinius*, a new Eocene insectivore. Am. Mus. Novitates **14**: 1-5. *2 figs.* 7 S 1921.
   [*Stehlinius uintensis* n. gen. & sp.]

1921h. Life in other worlds. Science **54** (**1394**): 239-241. 16 S 1921.

1921i. Fossil vertebrates and the Cretaceous-Tertiary problem. Amer. Jour. Sci. V. **2**: 209-227. 1 O 1921.
   [See 1921d. "Base of the true Eocene is the proper dividing line between Cretaceous and Tertiary."]

1921j. *Urus* and *Bison*. Natural History **21** (**6**): 598-606. *3 figs.* D 1921.

1921k. Why Palæontology? Natural History **21** (**6**): 639-641. *1 fig.* N-D 1921.

1921l. Notes. Palæontology [Jurassic of Cuba, Triassic of Texas, the ostrich dinosaur, and the Swabey mammoth skull.] Natural History **21** (**6**): 658-661. *2 figs.* N-D 1921.

1921m. Notes. [Snake Creek and Bolivian Mammals.] Natural History **21** (**6**): 661-662. N-D 1921. [Unsigned.]

1921n. A bibliographic list of the scientific publications of W. D. Matthew 1892-1921. Privately printed: Press of the New Era Printing Co. Lancaster, Pa. 31 D 1921.
   [Accompanied by a mimeographed "List of Omissions."]

1922a. Phyletic relations of the Lance vertebrates. Pan-Am. Geol. **37** (**1**): 68-69. F 1922.
   [Lower Paleocene should belong in Cretaceous; upper, in Tertiary. See 1921d and 1921i.]

1922b. "Errors and changes." In "The origin and evolution of the human dentition." By W. K. Gregory. Baltimore: Williams and Wilkins Co. xiii-xiv. 1922.
   [Corrections offered by Matthew after examination of specimens in European collections and elsewhere.]

1922c. [Discussion of] "Stratigraphy of the lower Oreodon beds of the South Dakota big badlands." (Abstract) by W. J. Sinclair with discussion by W. D. Matthew. Bull. Geol. Soc. Am. **33**: 156. 31 Mr 1922.

1922d. New light on the phylogeny of the Canidæ. (Abstract) Bull. Geol. Soc. Am. **33**: 214. 31 Mr 1922.
   [New material from the Lower Snake Creek is intermediate between Canidæ of the Oligocene and of the Pleistocene.]

1922e. Snake Creek fauna. (Abstract) Bull. Geol. Soc. Am. **33**: 215. 31 Mr 1922.

1922f. ————; **& Brown, Barnum.**
   The family Deinodontidae, with notice of a new genus from the Cretaceous of Alberta. Bull. Am. Mus. Nat. Hist. **46**: 367-385. *1 fig.* 31 My 1922.

1922g. A super-dreadnaught of the animal world, the armored dinosaur *Palæoscincus*. Natural History **22** (**4**): 333-342. *5 figs. 1 pl.* Jl-Au 1922.

1922h. ————; **& Andrews, Roy Chapman.**
   Gobi—a desert "wonder-house." Asia **22** (**12**): 859-862, 1000, 1002, and 1005. *8 figs.* D 1922.

1922i. **Osborn, Henry Fairfield; &** ————
   Explorations and researches of the Department of Vertebrate Palæontology. Preface to "Fossil Vertebrates in the American Museum of Natural History. **7**: iii-viii. 1918-1921." 1922.

1923a. Geological occurrence of the *Hesperopithecus* tooth. In "Notes on the type of *Hesperopithecus haroldcookii* Osborn." By W. K. Gregory and Milo Hellman. Am. Mus. Novitates **53**: 11-13. 6 Ja 1923.

1923b. ————; **& Granger, Walter.**
   Pliocene mammals of southern China. (Abstract) Bull. Geol. Soc. Am. **34**: 128. 30 Mr 1923.

1923c.   Stratigraphy of the Snake Creek fossil quarries and the correlation of the faunas. (Abstract) Bull. Geol. Soc. Am. **34**: 131.   30 Mr 1923.

1923d.   [Review.] "Grundzüge der Paläontologie. II Abt. Vertebrata." By K. A. von Zittel. Neuarbeitet von F. Broili und Max Schlosser. Science **58** (**1493**): 107-109.   10 Au 1923.

1923e.   Scientific names of Greek derivation. Nature **112**: 241.   18 Au 1923.

1923f.   Fossil Bones in the rock. The fossil quarry near Agate, Sioux County, Nebraska. Natural History **23** (**4**): 358-369.   *9 figs. 2 pl.* Jl-Au 1923.

1923g.   Recent progress and trends in vertebrate paleontology. (Presidential address to the Paleontological Society.) Bull. Geol. Soc. Am. **34** (**3**): 401-418.   30 S 1923. [Reprinted in Smithsonian Report for 1923, 273-289. 1925.]

1923h.   Fossil vertebrates. Natural History **23** (**5**): 520-522. S-O 1923.   [Unsigned.]

1923i.   ————; **& Brown, Barnum.**
Preliminary notices of skeletons and skulls of Deinodontidæ from the Cretaceous of Alberta. Am. Mus. Novitates **89**: 1-9.   *5 figs.*   11 O 1923.

1923j.   ————; **& Granger, Walter.**
New fossil mammals from the Pliocene of Sze-Chuan, China. Bull. Am. Mus. Nat. Hist. **48**: 563-598.   *27 figs.*   10 D 1923.
[An extensive fauna from cave deposits.]

1923k.   ————; **& Granger, Walter.**
The fauna of the Houldjin Gravels. Am. Mus. Novitates **97**: 1-6.   *6 figs.*   18 D 1923.
[Oligocene of Mongolia.]

1923l.   ————; **& Granger, Walter.**
The fauna of the Ardyn Obo formation. Am. Mus. Novitates **98**: 1-5.   18 D 1923.
[Faunal list. *Ardynia præcox* n. gen. & sp.]

1923m.   ————; **& Granger, Walter.**
New Bathyergidæ from the Oligocene of Mongolia. Am. Mus. Novitates **101**: 1-5. *4 figs.*   28 D 1923.
[Rodents from Hsanda Gol formation. *Tsaganomys altaicus* n. gen. & sp. Tsaganomyinæ new subfamily; *Cyclomylus lohensis* n. gen. & sp.]

1923n.   ————; **& Granger, Walter.**
Nine new rodents from the [Hsanda Gol] Oligocene of Mongolia. Am. Mus. Novitates **102**: 1-10.   *12 figs.*   31 D 1923.
[New genera: *Cricetops dormitor* n. gen. & sp., *Selenomys mimicus* n. gen. & sp., *Tataromys plicidens* n. gen. & sp., *Karakoromys decessus* n. gen. & sp., *Desmatolagus gobiensis* n. gen. & sp.]

1923o.   Dinosaur remains of the Red Deer Valley [Alberta, Canada]. Illustrated folder. 4 pages.
[Issued by the Canadian Pacific Railway.]

1924a.   Preliminary notes on the Mongolian faunas. Proc. Pan. Pac. Sci. Cong. (Melbourne, Australia, 1923) **1**: art. 66, sec. 7 (geology), div. 4 "Correlation of the Kainozoic formations of the Pacific region."   981-984.   1924.
[Faunal lists given. See 1926i.]

1924b.   ————; **& Granger, Walter.**
New Carnivora from the Tertiary of Mongolia. Am. Mus. Novitates **104**: 1-9. *7 figs.*   15 Ja 1924.
[Eocene and Oligocene; *Didymoconus colgatei* n. gen. & sp., *Amphicticeps shackelfordi* n. gen. & sp.]

1924c.   ————; **& Granger, Walter.**
New insectivores and ruminants from the Tertiary of Mongolia, with remarks on the correlation. Am. Mus. Novitates **105**: 1-7.   *3 figs.*   18 Ja 1924.
[*Tupaiodon morrisi* n. gen. & sp., *Palæoscaptor acridens* n. gen. & sp., *Eumeryx culminis* n. gen. & sp. Includes a list of the Hsanda Gol, Oligocene, fauna, 28 mammals of which about one-half are new genera.]

1924d. Notes. Vertebrate Palæontology. Welcomed home by a dinosaur. Replicas of *Baluchitherium* distributed. Fossil Birds from Nebraska. Natural History **24** (1): 118-119. *1 fig.* Ja-F 1924.

1924e. Tertiary terrestrial vertebrate horizons of North America. Bull. Geol. Soc. Am. **35**: 172. 30 Mr 1924.
[Title only.]

1924f. Memorial to George F. Matthew. Bull. Geol. Soc. Am. **35**: 181-182. 30 Mr 1924.
[Summarized remarks.]

1924g. Exploration for fossil remains in Mongolia. (Abstract) Bull. Geol. Soc. Am. **35**: 187-188. 30 Mr 1924.

1924h. Third contribution to the Snake Creek fauna. Bull. Am. Mus. Nat. Hist. **50**: 59-210. *63 figs.* 3 Jl 1924.
[Stratigraphy of the Snake Creek, faunal lists, faunal zones, correlations; descriptions of mammals, phylogeny of the Canidæ, validity of species of Equidæ, distribution of Equidæ, antlers of *Merycodus, Chelydrops stricta* n. gen. & sp. of soft-shell turtle.]

1924i. Fossil animals of India. Natural History **24** (4): 208-214. *7 figs.* Jl-Au 1924.

1924j. A new link in the ancestry of the horse. Am. Mus. Novitates **131**: 1-2. 23 S 1924.
[Abstract in Brit. Assoc. Adv. Sci., Rept. 92nd meeting **1925**: 380-381.]
[*Plesippus* n. gen., type *Equus simplicidens* Cope. *Pliohippus proversus* Merriam, referred.]

1924k. Notes. Extinct animals. Fossil horses from the Texas Pliocene. Natural History **24** (5): 629-631. S-O 1924.

1924l. [Notes on *Serridentinus* and "*Baluchitherium*."] In "*Serridentiinus* and *Baluchitherium*, Loh formation, Mongolia." By H. F. Osborn. Am. Mus. Novitates **148**: 5. 11 N 1924.

1924m. Notes. Vertebrate fossils. [A note on collections made by F. von Huene in South America.] Natural History **24** (6): 726-727. N-D 1924.

1924n. Correlation of the Tertiary formations of the Great Plains. Bull. Geol. Soc. Am. **35**: 743-754. 30 D 1924.
[Criteria, value of the Equidæ and other phyla, characteristic genera, Eocene sequences, table.]

1925a. Notes. [Barnum Brown. Santiago Roth. Fossil mammals from St. Petersburg, Florida.] Natural History **25** (1): 96-97. Ja-F 1925.

1925b. Blanco and associated formations of northern Texas. (Abstract) Bull. Geol. Soc. Am. **36** (1): 221-222. 30 Mr 1925.

1925c. Fossil mammal faunas of Florida. (Abstract) Bull. Geol. Soc. Am. **36** (1): 225. 30 Mr 1925.

1925d. The value of palæontology. Natural History **25** (2): 166-168. Mr-Ap 1925.

1925e. ————; & Barbour, Erwin Hinckley.
An American fossil giraffe, *Giraffa nebrascensis* n. sp. Bull. Neb. State Mus. 1 (4): 33-40. *figs. 14-20.* Ap 1925.
[Doubts are expressed as to the nature of this specimen. It consists of two worn cheek teeth.]

1925f. ————; & Granger, Walter.
Fauna and correlation of the Gashato formation of Mongolia. Am. Mus. Novitates **189**: 1-12. *14 figs.* 7 O 1925.
[Age, somewhere between Lower Cretaceous and Upper Eocene. Notoungulate, multituberculate, rodent, creodont and (?) marsupial present. *Palæostylops iturus* n. gen. & sp., *Baënomys ambiguus* n. gen. & sp., *Prionessus lucifer* n. gen. & sp., *Eurymylus laticeps* n. gen. & sp., *Phenacolophus fallax* n. gen. & sp., *Hyracolestes ermineus* n. gen. & sp., *Sarcodon pygmæus* n. gen. & sp.]

1925g. ————; & Granger, Walter.

New creodonts and rodents from the Ardyn Obo formation of Mongolia. Am. Mus. Novitates **193**: 1-7. *9 figs.* 27 O 1925.

[*Ardynictis furunculus* n. gen. & sp., *Ardynomys olseni* n. gen. & sp.]

1925h.   ————; **& Granger, Walter.**

New ungulates from the Ardyn Obo formation of Mongolia with faunal list and remarks on correlation. Am. Mus. Novitates **195**: 1-12. *13 figs.* 19 N 1925.

[*Paracolodon curtus* n. gen. & sp., *Miomeryx altaicus* n. gen. & sp.]

1925i.   ————; **& Granger, Walter.**

New mammals from the Shara Murun Eocene of Mongolia. Am. Mus. Novitates **196**: 1-11. *10 figs.* 20 N 1925.

[*Olsenia mira* n. gen. & sp., *Deperetella cristata* n. gen. & sp., *Cænolophus promissus* n. gen. & sp., *Archæomeryx aptatus* n. gen. & sp.—"an approximate ancestral type for the Pecora."]

1925j.   ————; **& Granger, Walter.**

New mammals from the Irdin Manha Eocene of Mongolia. Am. Mus. Novitates **198**: 1-12. *10 figs.* 21 N 1925.

[*Gobiohyus orientalis* n. gen. & sp.]

1925k.   ————; **& Granger, Walter.**

The smaller perissodactyls of the Irdin Manha formation, Eocene of Mongolia. Am. Mus. Novitates **199**: 1-9. *figs. 9.* 23 N 1925.

[Lophialetinae n. subfam., *Teleolophus medius* n. gen. & sp., *Lophialetes expeditus* n. gen. & sp.]

1925l.   A dissenting opinion. In, "The origin of species. II Distinctions between rectigradations and allometrons." By H. F. Osborn. Proc. Nat. Acad. Sci. **11** (**12**): 751. D 1925.

1925m.   ————; **Madison, Harold Lester; and others.**

Code of ethics for museum workers. Being the report of a committee of the Am. Assoc. of Museums adopted unanimously at the twentieth meeting of the association 1925. The Am. Assoc. of Museums, New York **1925**: 8 pages. 1925.

[A preliminary draft of this code entitled "A tentative code of museum ethics" was drawn up by Harold L. Madison, Chairman of Committee on Ethics, and printed for discussion at the Twentieth Annual Meeting held at St. Louis, Missouri. May 17-21, 1925.]

1925n.   **Osborn, Henry Fairfield; &** ————

Explorations and researches of the Department of Vertebrate Palæontology. Preface to "Fossil Vertebrates in the American Museum of Natural History **8**: iii-x. 1922-1924." 1925.

1926a.   Vertebrate Palæontology. American Year Book. **1925**: 933-934. 1926.

1926b.   ————; **& Granger, Walter.**

Two new perissodactyls from the Arshanto Eocene of Mongolia. Am. Mus. Novitates **208**: 1-5. *5 figs.* 16 F 1926.

[*Schlosseria magister* n. gen. & sp., *Teilhardia pretiosa* n. gen. & sp.]

1926c.   [Note on rhinoceros remains from the Ricardo.] In "New canid and rhinocerotid remains from the Ricardo Pliocene of the Mohave Desert, California." By Chester Stock & Eustace L. Furlong. Univ. Calif. Publ. Geol. **16** (**2**): 50. 16 Mr 1926.

1926d.   **Osborn, Henry Fairfield; &** ————

Summary of discoveries in vertebrate paleontology in three seasons of field explorations in Mongolia. (1922, 1923, and 1925.) (Abstract) Bull. Geol. Soc. Am. **37**: 157-158. 30 Mr 1926. [See also 228.]

[Discovery of new Cretaceous and Tertiary faunas.]

1926e.   ————; **Holmes, W. W. & Megathlin, G. R.**

Pleistocene fauna from Seminole, Pinellas County, Florida. (Title only.) Bull. Geol. Soc. Am. **37**: 244. 30 Mr 1926.

1926f. The evolution of the horse. A record and its interpretation. Quart. Rev. Biol.
        1 (2): 139-185. *27 figs. 1 pl.* Ap 1926.
        [Critical summary of the evidence with many new facts and interpretations. Re-
        duction of the phalanges, changes in carpus and tarsus, lower limbs, skull, and
        teeth. Morphological stages and the stratigraphic sequence in North America
        and Europe. Tables showing geologic and geographic distribution, and phy-
        logeny of Perissodactyla. Evidences of evolution. Bibliography.]
1926g. On a new primitive deer and two traguloid genera from the Lower Miocene of Ne-
        braska. Am. Mus. Novitates **215**: 1-8. *3 figs.* 10 My 1926.
        [*Machæromeryx tragulus* n. gen. & sp.]
1926h. Early days of fossil hunting in the High Plains. Natural History **26** (5): 449-454.
        3 pls. S-O 1926.
1926i. Some comments on the characters of the fossil vertebrate faunas of Mongolia.
        (Abstract) Bull. Geol. Soc. China **5** (1): 5-9. 1926.
        [Discusses positions of three Cretaceous and five Tertiary faunas. Abstract by
        Matthew in Biol. Abstracts **2** (3-5): abstract (7160). Mr-My 1928.]
1926j. The most significant fossil finds of the Mongolian expedition. In "Important re-
        sults of the Central Asiatic expeditions." By C. P. Berkey, and others. Natural
        History **26** (5): 532-534. *5 figs. 9 pls.* S-O 1926.
1926k. Relations of vertebrates to sediments and sedimentary environments. In "Treatise
        on Sedimentation." By W. H. Twenhofel. 1st ed. 146-150.
        [Also in 2nd ed. 183-186.]
1927a. Notes. The Siwalik beds. Natural History **27** (2): 187. Mr-Ap 1927.
1927b. [Various items] in "Geology of Mongolia." By C. P. Berkey and F. K. Morris.
        Nat. Hist. of Central Asia; Amer. Mus. Nat. Hist. **2**: i-xxxi + 1-475. *161 figs.
        44 pls. 6 maps.*
        ["To Dr. W. D. Matthew, whose wide knowledge of palæontology and Cenozoic
        history make him an authority, we have referred many questions particularly
        as to the correlation of a number of the Tertiary formations." p. vii.]
1927c. A ballad of *Pithecanthropus*.
        [Privately printed: 4. New York.]
1928a. The evolution of the mammals in the Eocene. Proc. Zool. Soc. London **1927**:
        947-985. *16 figs.* 12 Ja 1928.
        [Progressive differentiation from Triassic to Tertiary; Cretaceous placentals; Creta-
        ceous character of the Paleocene mammals; conspectus of Paleocene and Lower
        Eocene forms; evolution through the Eocene; phylogeny of Carnivora, Peris-
        sodactyla, Artiodactyla. Theory of arboreal Eocene types, becoming diversified
        and specialized with development of brain and marked features in their evolu-
        tion. Abstract by L. S. Russell in Biol. Abstracts. (2968). Ja 1932.]
1928b. *Xenotherium* an edentate. Jour. Mam. **9** (1): 70-71. F 1928.
        [Lower Oligocene of Montana.]
1928c. ————; & Siegfriedt, J. C. F.
        New fossil mammals from the Fort Union of Montana. (Abstract) Bull. Geol.
        Soc. Am. **39**: 300. 30 Mr 1928.
        [Finding of marsupials suggests North American origin for cænolestids. MS. of
        entire paper in Matthew library.]
1928d. Affinities of some new paleocene mammals from Fort Union beds. (Abstract)
        Pan-Am. Geol. **49** (4): 319. My 1928.
        [Also in Bull. Geol. Soc. Am. **40**: 258. 30 Mr 1928. (Title only.)]
        [MS. of entire paper in Matthew library.]
1928e. Correlación de las formaciones Cenozoicas en la Argentina.—Carta del Dr. Wil-
        liam D. Matthew a Alfredo Castellanos. New York, 11 de Octobre de 1923.
        Tradduccion del ingles. [Being the "Appendix" to Castellanos "Notas criticas

sobre el puelchense de las sedimentos neogenos de la Argentina." Rev. Univ. Nac. Cordova **15** (5-6): 52-54.] Jl-Au 1928.

[Discusses ground sloth distribution and occurrences in N. & S. Am. Also *Glyptodon,* carnivores, and equids.]

1928f.   Dinosaurs and rickets. Sci. News Letter 15 Au 1928.

[Criticism of article by Marshall in S. N. L. Au 1928.]

1928g.   Synopsis lectures in Paleontology 1. Outline and general principles of the history of life. Univ. Calif. Syllabus Series **213**: 1-253. *21 figs.* Au 1928.

[For 2nd ed. see Matthew 1930h.]

[Used in elementary classes at the University of California and elsewhere. A masterpiece of simplified presentation. Many of the figures were drawn by Matthew.]

1928h.   ——————; **Granger, Walter; & Simpson, George Gaylord.**

Paleocene multituberculates from Mongolia. Am. Mus. Novitates **331**: 1-4. *3 figs.* 30 O 1928.

[*Sphenopsalis nobilis* n. gen. & sp.; affinities of *Prionessus* and *Sphenopsalis.*]

1928i.   [Review.] "Die Saugethiere, Einführung in die Anatomie und Systematik der recenten und fossilen Mammalia." By Max Weber. (2nd edition. Edited by O. Abel.) Science **68** (1766): 429-430. 2 N 1928.

1928j.   The ape-man of Java. Natural History **28** (6): 577-588. *7 figs.* N-D 1928.

[A popularized word-picture of the life and environment of Pleistocene man. Reprints were distributed to the elementary classes in Paleontology at the University of California in 1928.]

1929a.   A new and remarkable hedgehog from the later Tertiary of Nevada. Univ. Calif. Publ. Geol. **18** (4): 93-102. *pls. 7, 8.* 29 Ja 1929. [*Metechinus nevadensis* n. gen. & sp.]

1929b.   Critical observations upon Siwalik mammals (exclusive of Proboscidea). Bull. Am. Mus. Nat. Hist. **56**: 437-560. *55 figs.* 9 F 1929.

[Result of re-examination of type collections. Correlation of faunas with Europe and N. Amer.; comments on faunal lists; review of collections in Calcutta and British Museum; phylogeny of Ursidæ; revision of the Machærodontinæ; notes on Siwalik rhinoceroses; notes on the Pondaung fauna of Burma; critical observations on Siwalik perissodactyls in British Museum; classification of the Chalicotheriidæ; phylogeny of Giraffidæ; notes on Camelidæ, Hippopotamidæ and Rodentia. Abstract by Matthew in Biol. Abstracts (2967) Ja 1932.]

1929c.   On the phylogeny of horses, dogs and cats. Science **69** (1793): 494-496. 10 My 1929.

[An answer to Austin Clark's theory of evolution. Abstract by Matthew in Biol. Abstracts (26276). O 1930.]

1929d.   Preoccupied names. Jour. Mam. **10** (2): 171. My 1929.

[*Stehlinius* preocc. becomes *Stehlinella. Palæolestes* preocc. becomes *Prodiacodon.*]

1929e.   Reclassification of the artiodactyl families. Bull. Geol. Soc. Am. **40**: 403-408. 30 Je 1929.

["The customary grouping of the families of Artiodactyla is unsatisfactory as regards the extinct forms of the older Tertiary. The following arrangement is suggested;" Palæodonta, new suborder (Dichobunidæ and Entelodontidæ), Hyodonta, new suborder (dicotyls, pigs and hippopotamuses) Ancodonta, new suborder (anthracotheres, anoplotheres, cænotheres and oreodonts), Tylopoda (xiphodonts and camels), Pecora (true ruminants). Abstract by Matthew in Biol. Abstracts (26277). O 1930.]

1929f.   The phylum in zoology and paleontology. Science **70** (1806): 142-143. 9 Au 1929.

[Continuation of the discussion with Austin Clark. Phylum, as used in Paleontology designates what Clark has termed "linear evolution involving a time element."]

1929g. ——————; **Granger, Walter; & Simpson, George Gaylord.**
Additions to the fauna of the Gashato formation of Mongolia. Am. Mus. Novitates **376**: 1-12. *11 figs.* 7 O 1929.
[Revised faunal list included. *Praolestes nanus* n. gen. & sp., *Pseudictops lophiodon* n. gen. & sp., Eurymylidæ n. fam., *Opisthopsalis vetus* n. gen. & sp., *Prodinoceras martyr* n. gen. & sp.]

1929h. [Paleontological articles for] Encyclopædia Britannica, 14th edition. London and New York. Encyclopædia Britannica, Ltd. 24 Vols. 1929.
[Amblypoda **1**: 740; Carnivora: extinct forms **4**: 900-901; Creodonta **6**: 667, 668; Equidæ **8**: 670-672; *Glyptodon* **10**: 449; horse: fossil horses **11**: 757; *Machærodus* **14**: 574; *Megatherium* **15**: 214-215; *Mylodon* **16**: 42; *Palæotherium* **17**: 112; *Phenacodus* **17**: 698; Toxodontia **22**: 337; Tylopoda **22**: 641-642.]

1929i. [Note on distribution of Pleistocene mammals.] In "The pre-Illinoian Pleistocene geology of Iowa." By G. F. Kay and E. T. Apfel. Iowa Geol. Survey Ann. Rept. **34**: [122 and 194].

1930a. Boundary between the Pliocene and the Pleistocene. (Abstract) Bull. Geol. Soc. Am. **41** (I): 155. 31 M 1930.
[By title. Discussed by Dr. B. L. Clark.]

1930b. We discover a mastodon. Standard Oil Bulletin (Standard Oil Company of Calif.) **18**: 12-14. *3 figs.* My 1930.
[*Pliomastodon vexillarius* n. sp. This is the original description. See 1930j.]

1930c. ——————; **& Stirton, Ruben Arthur.**
Osteology and affinities of *Borophagus*. Univ. Calif. Publ. Geol. **19** (7): 171-216. *2 figs. pls. 21-34.* 9 My 1930.
[First paper on the Hemphill faunas, Pliocene of Texas. Includes preliminary faunal list of the Coffee Ranch Quarry at Miami, and phylogeny of the hyaenoid Canidæ.]

1930d. The phylogeny of dogs. Jour. Mam. **2** (2): 117-138. *3 figs.* My 1930.
[Canis derived from *Tephrocyon*, *Cyon* from *Temnocyon*, *Procyon* from *Phlaocyon*, *Ursus* through *Arctotherium* from *Hemicyon*.]

1930e. Credit or responsibility in scientific research. Science **71** (**1852**): 662-663. 27 Je 1930.

1930f. Range and limitations of species as seen in fossil faunas. Bull. Geol. Soc. Am. **41**: 271-274. 30 Je 1930.
[Abstract in Bull. Geol. Soc. Am. **41** (I): 210-211. 30 Mr 1930.]
["We should not . . . expect to find two or more closely related species living together at the same time, within the same area, and with the same habits, causing their remains to be preserved together in the same quarry."]

1930g. The dispersal of land animals. Scientia **48**: 33-42. Jl 1930.
[Continues the thesis advanced in "Climate and Evolution." See 1915d.]

1930h. Synopsis of lectures in Paleontology 1. Outline and general principles of the history of life. Univ. Calif. Syllabus Series **230**: 1-276. *26 figs.* Au 1930.
[2nd edition of 1928g.]

1930i. The pattern of evolution: a criticism of Doctor Austin Clark's thesis. Scientific American **143**: 192-196. *5 figs.* S 1930.
[See Austin Clark, Scientific American, **143**: 104 Au 1930.]

1930j. A Pliocene mastodon skull from California, *Pliomastodon vexillarius* n. sp. Univ. Calif. Publ. Geol. **19** (16): 335-348. *2 figs. pls. 41-44.* 26 N 1930.
[For original description see Matthew 1930b.]

1930k. ——————; **& Stirton, R. A.**
Equidæ from the Pliocene of Texas. Univ. Calif. Publ. Geol. **19** (17): 349-396. *pls. 45-58.* 29 N 1930.

[Continuation of study of faunas of the Hemphill beds. *Calippus* n. subgen., type *Protohippus placidus* Leidy. Distinctive characters of later Tertiary horses summarized.]

1931a. Critical observation on the phylogeny of the rhinoceroses. Univ. Calif. Publ. Geol. **20** (1): 1-9. *2 figs.* 23 Ja 1931.

[*Dicerorhinus* of Sumatra most primitive of living forms. Presence or absence of lower tusks, presence or absence of nasal and frontal horns are characters used to separate the main phyla. Modern rhinos regarded as relicts of the Tertiary forms.]

[Abstract Bull. Geol. Soc. Am. **42**: 366-367.]

1931b. [Abstract.] "The evolution of Earth and Man." By G. A. Baitsell, and others. Yale Univ. Press 1929. Biol. Abstracts **5** (10): abstract (22669). O 1931.

1931c. Genera and new species of ground sloths from the Pleistocene of Cuba. Am. Mus. Novitates **511**: 1-5. 16 D 1931.

[With prefatory note by Walter Granger. See 1915f, 1918h, 1918l, and 1919g. Three new species of *Mesocnus* and *Miocnus* are described.]

1932a. A review of the rhinoceroses with a description of *Aphelops* material from the Pliocene of Texas. (A posthumous paper edited by R. A. Stirton.) Univ. Calif. Publ. Geol. **20** (12): 411-482. *12 figs. pls. 61-79.* 26 F 1932.

[Characters of *Aphelops, Peraceras, Teleoceras;* comparisons with *Chilotherium* and other Old World genera; discussion of principles of dispersal of phyla.]

1932b. New fossil mammals from the Snake Creek quarries. Am. Mus. Novitates **540**: 1-8. *7 figs.* 16 Je 1932.

1932c. A skeleton of *Merycoidodon gracilis* and its adaptive significance. Univ. Calif. Publ. Geol. **22** (5): 13-30. *pls. 2, 3.* 12 Jl 1932.

[Species of *Merycoidodon;* adaptive significance of skull, habits, environment and causes of extinction of oreodonts.]

1932d. [Correspondence on the Hemphill County Pliocene faunas of Texas.] In "The geology of Hemphill County, Texas." By L. C. Reed, and O. M. Longnecker, Jr. Univ. Texas Bull. **3231**: 69-70. 1932.

[Discusses the age of the beds near Canadian and the origin of the deposits.]

1933a. ——————; & **Mook, Charles Craig.**

New fossil mammals from the Deep River beds of Montana. Am. Mus. Novitates **601**: 1-7. *2 figs.* 22 Mr 1933.

[Pt. I. Occurrence, by C. C. Mook. Pt. II. Descriptions, by William Diller Matthew. *Brachyerix macrotis* n. sp. hedgehog. *Sciurus angusticeps* n. sp. squirrel.]

1937a. Paleocene faunas of the San Juan Basin, New Mexico. Trans. Am. Philos. Soc. **30**: i-viii + 1-510, *85 figs. pls. 1-55.*

Additions to William Diller Matthew bibliography:

1934. ———; with Edwin H. Colbert. "A phylogenetic chart of the Artiodactyla." *Journal of Mammalogy* 15: 207–209.

1943. "Relationships of the orders of mammals." Edited and annotated by George Gaylord Simpson. *Journal of Mammalogy* 24: 304–311.

1960. ———; with J. R. Macdonald. "Two new species of *Oxydactylus* from the middle Miocene Rosebud formation in western South Dakota." *American Museum Novitates* 2003: 1–7.

INDEX

Abel, Othenio, 190

Academy of Natural Sciences (Philadelphia), 45

Acquired characteristics, 182

Adaptive radiation, 181, 236

Adirondacks, 39–40

Africa, 177, 179, 192

Agassiz, Louis, 23

Agate Fossil Beds (Nebraska): origin of fossil deposits at, 161–162; described, 144, 151, 162; fieldwork at, 130, 144–150, 158, 160–161, 173; importance of, xi, 144–147, 149, 160–162; as national monument, 152, 173; permanent camp at, 151–152, 153

Agate Springs Ranch, 151, 172–173

Alticamelus, 143

Amblypods, 138

American Museum Journal, 175

American Museum of Natural History (New York), 1, 48, 188, 216, 218, 219, 223, 225, 230, 232; architecture of, 222; Carnegie Museum and, 70, 86–87, 124, 148–150, 160–161; differences at, 113, 116–123; exhibitions, 51, 57, 64, 95, 124–125, 161, 173, 189; funding, 52, 92, 127, 170–171, 192–193; Library, 220; Mosasaur at, 64–65; *Plateosaurus* at, 189; preparation laboratory, 124; salaries at, 46, 49, 58–60; taxidermy department, 125–126; *Tyrannosaurus* at, 95; vertebrate paleontology department, 38–39, 43, 59, 114, 116–117, 121–126, 129, 234; working environment, 123

American Museum of Natural History Bulletins, 52, 135, 139, 173–174, 203–204, 229

American Museum of Natural History field expeditions: Central Asia, 127, 130, 192–201, 228–229; Colorado, 65–67, 141; Kansas, 58–59, 62–65, 67; Montana, 83; Nebraska, 130, 144, 147–158, 160–172; North Carolina, 46–47, 117; South Dakota, 91, 96–103, 129–131, 141, 143–144; Wyoming, 63–65, 67–70, 84, 126, 127, 129

American Museum of Natural History fossil collections: Agate, 160–162; Bridger Basin, 136; Cope, 51–53, 61–62, 135; permanent, 58; Siwalik, 194, 204; size of, 43, 87, 129; Snake Creek, 158–159; Warren, 92–93

Home Guard, 184–185

Hominids, 133–134, 175, 192, 193, 203, 205–209, 226

Homo erectus, see Pithecanthropus

Horses: Eocene, 134, 137, 139; evolution of, 120–122, 175, 176, 180–181, 216, 226; grass and, 180–181; Miocene, 141; Oligocene, 141, 143; at Sheep Creek, 168; at Snake Creek, 156–158, 166; in Texas, 230

Howard, Hildegarde, 228

Hudson River, 55, 94, 105, 110, 115, 118, 127, 198

Hunt, Robert M., Jr., 146, 161–162, "The Agate Hills" (1984), 149, 150, 162*n*

Huxley, Thomas Henry, 43, 44

Hyracotherium, 139

Ichthyology, 182

India, 177, 193–194, 200, 202–205, 229

Indian Geological Survey, 202

Indian Museum (Calcutta), 194, 202–205

Influenza epidemic (1917–18), 110, 185

Inland lake theory, 142

Insectivores, 61, 105, 133–134, 136–137, 139, 181–182, 236

International Geological Congress (Paris, 1900), 71, 74–76

Invergowrie (Scotland), 2

Ireland, 3

Italy, 187, 189–190

Jack, William Brydone, 23, 28

Jacob, Edwin, 22–23

Jarrell, Richard A., 29

Java, 203, 205–213

Java Geological Survey, 206–208

Jorgensen, Chris, 233

Jurassic: dinosaurs, 67, 126, 133; mammals, 133; Wyoming, 67, 126, 128–129

Kaisen, Peter, 69, 70, 124

Kalgan (China), 195–197, 199

Kangaroos, 180

Kansas, 58–59, 62–65, 67

Kansas University, 64, 65

Keith, Arthur, 212

Kellogg, Remington, 228

Kemp, James Furman, 39–40, 44, 46, 50, 219

Kennebecasis River (New Brunswick), 4, 11, 17, 26, 88–89, 233

King, Clarence, 142

King crab, 17

King's College (New Brunswick), *see* New Brunswick, University of

King's College (New York City), *see* Columbia University

Kinne, A. M., 233

Kipling, Rudyard, 194

Knight, Charles R., 124–125

Lake Champlain, 39

Lamarck, Jean Baptiste de, 182

Land bridges, 179–181

Latham, Marion E., 176

Laurasia, 178

Laval University, 6

Lawson, Andrew C., 233

LeConte, Joseph, 215

Lee, Bess, 77, 79, 80

Lee, Charles, 77

Lee, Edward, 77

Lee, Emma, 77, 79, 85, 94, 234, 235

Lee, James Edgar, 77

Lee, Jane, *see* Pomeroy, Jane Eliza

Lee, Kate, *see* Matthew, Kate Lee

Lee, Mary (Mame), 77, 107, 112, 216

Lee, Montague (Mont), 77

Lee, Pomeroy (Roy), 77

Lee, Robert E., x

Leidy, Joseph, 136

Life zone survey, 119

Limulids, 17

Little Medicine River (Wyoming), 68–69

Loess, 167

London, 76, 190–191, 194, 214, 217–219, 229

Loomis, Frederick B., 145, 148, 150

67, 141–143, 181; "Outline and General Principles of the History of Life" (1928), 226; "Paleocene Faunas of the San Juan Basin" (1937), 135, 140, 182, 229; "The Pattern of Evolution" (1930), 120*n*; "A Pliocene Fauna from Western Nebraska" (1909), 173; "A Provisional Classification of the Fresh-water Tertiary of the West" (1899), 52; "A Revision of the Lower Eocene Wasatch and Wind River Faunas: Part I—Order Ferae (Carnivora), Suborder Creodonta" (1915), 139*n*; "A Revision of the Puerco Fauna" (1897), 52, 60–61, 135; *The Science-History of the Universe, Volume VI: Zoology and Botany* (1909), 176; "Third Contribution to the Snake Creek Fauna" (1924), 173–174, 182

Matthew, William Pomeroy (Roy/Bill) (son), ix, 186, 194, 220, 230; birth of, 107; in California, 221, 235; in Connecticut, 234

Maybeck, Bernard, 223

McGill University, 59–60

McGowan, Samuel, xii

Mechanics Institute Museum (Saint John), 5

Mediterranean Sea, 177

Mendel, Gregor Johann, 120

Menke, H. W., 67

Menoceras, 147

Mentor, 175

Merriam, John C., 214–215, 228

Merychippus isonesus primus, 122

Merychippus isonesus quintus, 122

Merycochoerus, 100

Mesozoic era: birds, 134; mammals, 133, 178; plants, 178; reptiles, 129, 132–133

Metallurgy, 36–37

Metropolitan Museum of Art (New York), 38

Miacids, 139

Mice, 141

Microclaenodon assureus, 61

Microconodon, 45

Middlebury (Vermont), 13, 128

Middlebury College, 128

Migration, 177–181

Miller, Loye, 228

Miller, Paul 147

Miller, Randall F., 28

Mineralogy, 34, 41

Mining, 30, 31, 36–37, 44, 114

Minor, Kate, *see* Matthew, Kate Lee

Minor, Ralph, ix, 234–235

Miocene, 129, 142, 171; mammals, 65, 141, 146–147; Snake Creek, 154

"Missing link," 205, 206, 211, 213

Mongolia, 127, 130, 140, 198, 200; dinosaurs in, 193; early mammals in, 193, 228–229, 236

Monotremes, 133

Montana, 83

Morgan, J. P., 38, 52, 92

Morgan, Thomas Hunt, 120

Moropus, 147, 160

Mosasaur, 62–65

Moses, Alfred Joseph, 34

Multituberculates, 133

Murchison Medal, 6

National Academy of Science, 1–2

National Park System, 152, 173

National Science Foundation, 170, 192

Natural History, 161, 175

Natural History Society (Saint John), 5–6

Natural selection, 119, 120, 180–183

Nebraska, 129, 130, 144–174, 221

Nebraska, University of, 145, 146, 147, 150, 151, 156, 161

Nelson, Nels, 200

Nevada, 230

Newberry, John Strong, 37

New Brunswick (Canada), 4–6, 8, 11, 23, 39, 44; *see also* Saint John

New Brunswick College, *see* New Brunswick, University of

New Brunswick Museum, 6, 18, 28

New Brunswick, University of, 6, 17;